SOUND FOUNDATIONS

A PRACTICAL MATHEMATICS SIMULATION

ROBERT K. GERVER
RICHARD J. SGROI
North Shore High School
Glen Head, New York

ISBN: 0-538-13121-7

5 6 7 8 9 H 4 3 2 1

Printed in the United States of America

Published by

M12 **SOUTH-WESTERN PUBLISHING CO.**

CINCINNATI WEST CHICAGO, IL DALLAS LIVERMORE, CA

PREFACE

To the Student

Mathematical skills will play an important role throughout your life. Many decisions will require accurate problem-solving skills and computational skills. Learning these mathematical skills can be both interesting and exciting.

SOUND FOUNDATIONS – *A Practical Mathematics Simulation* is designed to strengthen your basic math skills. SOUND FOUNDATIONS will show you how math is used in the rock music industry. Read the Introduction on pages 1–3 carefully. Good luck as you manage a band in New York City and build SOUND mathematical FOUNDATIONS!

To the Teacher

SOUND FOUNDATIONS – *A Practical Mathematics Simulation* is an innovative program designed to strengthen students' fundamental math skills. Its unique "mathemusical" theme motivates the learning of basic math skills and fosters a challenging and thrilling educational experience. Students are intrigued by the glitter and glamour of the rock music industry. The fact that mathematics is used so extensively in the music industry is quite a revelation to them. SOUND FOUNDATIONS taps the students' enthusiasm and directs it toward achieving mathematical competency.

Each student will assume the role of band manager and lead a musical group from conception to success. While managing the bands, the students will develop a proficiency with basic mathematics, such as whole numbers, fractions, decimals, percents, ruler skills, area, perimeter, volume, circumference, graphs, signed numbers, and formulas, backed by a motivating musical accompaniment.

Simulation Objectives

SOUND FOUNDATIONS provides students with a comprehensive, motivating basic mathematics skills curriculum. SOUND FOUNDATIONS will help students improve their proficiency and self-confidence in mathematics while giving them an inside look into the world of show business. The activities in SOUND FOUNDATIONS bring math and music together in a practical harmony. The music industry theme is designed to create enthusiasm about math and its applications. The experience of playing the role of band manager should help students become more conscientious and independent in the math class.

SOUND FOUNDATIONS can be implemented in both the junior and senior high schools. It is suitable as a refresher course in basic math skills and as a preparatory course for competency tests in mathematics. SOUND FOUNDATIONS can be used in conjunction with any math textbook.

Special Features of the Simulation

This *text-workbook* is divided into ten units called milestones. The ten milestones serve as a musical backdrop as students develop basic math skills. Each milestone represents a key developmental stage in the band's life. The students study specific mathematical topics encountered along the road to success. The text-workbook features activities and milestone forms. Each activity gives the student an opportunity to practice a newly acquired mathematical skill by applying the skill in different aspects of the music business. Each milestone form simulates the activities of a rock band's rise to success. These forms involve the purchase of a truck and band equipment, rehearsals, dance clubs, record sales, radio airplay, and concert tours.

A comprehensive *teacher's manual* accompanies the simulation. Its explicit format makes it an indispensable tool in the mathematics classroom. The teacher's manual includes a detailed overview to the simulation, an explanation of the teacher's responsibilities in the simulation, lesson plans for the entire year, and solutions to all activities, quizzes, and tests. The teacher's manual explains each mathematical and musical aspect of SOUND FOUNDATIONS. No business or music knowledge is necessary to use the simulation.

An envelope of *teacher's supplies* includes 111 daily quiz masters, 23 transparency masters, 8 test masters, 23 forms masters, and 5 sets of cards.

About the Authors

Robert Gerver and Richard Sgroi created SOUND FOUNDATIONS in the summer of 1980 for use in their mathematics course at North Shore High School, Glen Head, New York. SOUND FOUNDATIONS has been successful in helping students meet state requirements in mathematics competency.

Both authors are doctoral candidates in the Mathematics Education Ph.D. program at New York University. In addition, they have taught mathematics and math education at Long Island University, Adelphi University, Dowling College, State University of New York, and an institute for gifted and talented students on Long Island. They have spoken at numerous national and regional conferences and have published articles in mathematics journals.

Acknowledgments

The authors would like to thank Barbara Gerver for her countless hours of typing of the SOUND FOUNDATIONS package. Robert Gerver dedicates SOUND FOUNDATIONS – *A Practical Mathematics Simulation* to Jules and Ruth Gerver for a lifetime of support and encouragement. Richard Sgroi dedicates SOUND FOUNDATIONS – *A Practical Mathematics Simulation* to Laura Sgroi for her enthusiasm, supportive critiques, and patience throughout the various stages of the project.

Robert Gerver
Richard Sgroi

CONTENTS

PHOTO CREDITS

For permission to reproduce the photographs on the pages indicated, acknowledgment is made to the following:

Page 28 Gibson, Inc.
Slingerland/Deagan
Zildjian
Slingerland/Deagan

Page 55 Slingerland/Deagan

INTRODUCTION

THE HISTORY OF ROCK 'N' ROLL

Rock 'n' roll music began in the mid – 1950s in the southern part of the United States. Rock 'n' roll was a new type of music, and many people were shocked by its unusual sound. It was a mixture of country, folk, and blues music with a heavy, repeated beat. With the arrival of the Beatles in the early 1960s, rock music became a respected form of entertainment. Since that time, rock music has grown tremendously. In addition to live entertainment, rock music now has an important place in television, radio, newspapers, films, magazines, and advertising. Rock 'n' roll is more than music – it's a billion dollar business, and you'll become part of this business in

SOUND FOUNDATIONS

A great deal of work and planning must be done before a band can succeed. Most of this work is behind the scenes, and most people never see it. Did you ever wonder who plans a band's concerts and albums? The musicians in the band have a manager to keep track of all the behind-the-scenes details, and

YOU ARE NOW A BAND MANAGER! GOOD LUCK!!

DUTIES OF THE BAND MANAGER

The band manager in SOUND FOUNDATIONS – *A Practical Mathematics Simulation* makes most of the musical and business decisions for the band. Many decisions must be made about buying equipment, rehearsing, making records, advertising, and playing live concerts. Mathematics plays an important role in much of the band manager's work. Keeping track of income and expenses will be your most important job.

KEY PEOPLE

Many people work behind the scenes of a successful band. In addition to the band members, you will be working with several other people. Each of these people has a specific job.

- ROADIES Each band in SOUND FOUNDATIONS has three roadies. Roadies are responsible for setting up and transporting all the band's musical equipment.
- SOUND CREW The sound crew adjusts the volume and tone levels on all the microphones and musical instruments in rehearsals and concerts.

- LIGHT CREW The light crew sets up all light fixtures and special lighting effects for live performances. They conduct the light show during a concert.
- AGENT The agent finds jobs for the band. The agent must be in touch with dance clubs, concert halls, and producers.

All these key people must be present at all live performances. They receive a share of the profits as their wages. As band manager, you will compute their salaries.

A Top Quality Band Equals Success

What makes a band successful?

- great equipment
- many hours of rehearsal
- plenty of publicity
- a reliable manager. The band manager must be punctual, dependable, and an accurate record keeper.

These qualities contribute to the success of the band. A careful investment of time and money will pay off with gold records and sell-out concerts. In SOUND FOUNDATIONS, your band's quality is measured by quality points.

Quality Points

A *quality point* is a numerical rating that is placed on the quality of the equipment you purchase, the time you rehearse, and the publicity you purchase. Quality points are a measure of your effort to create a successful band. Your teacher will assign the quality points, and you will keep an accurate record of your quality points.

Location of Your Band

Your band is based in New York City. New York City, nicknamed the "Big Apple," has many concert halls, recording studios, radio stations, TV studios, and music stores. It is the best place to start a rock 'n' roll band. You will be living and working in New York City!

Improving Your Math Skills

SOUND FOUNDATIONS will help you improve your mathematics skills while giving you an inside look into the glitter and glamour of the music industry. The activities in SOUND FOUNDATIONS bring math and music together in a practical harmony.

The road to success for you in managing your own band is marked by ten milestones:

Milestone 1 – Band Formation
Milestone 2 – Equipment Purchase
Milestone 3 – Rehearsal
Milestone 4 – Dance Clubs
Milestone 5 – Record Sales
Milestone 6 – Airplay
Milestone 7 – Publicity
Milestone 8 – Local Concerts
Milestone 9 – Away Concerts
Milestone 10 – National Tour

Along this road to success, you will sharpen your math skills with fractions, decimals, percents, geometry, and more! You will need these math skills on days when you are taking care of band business.

Please note (♪): You are a very important business person now. You must do your work neatly and carefully. And above all, remember that

MATHEMATICS IS THE

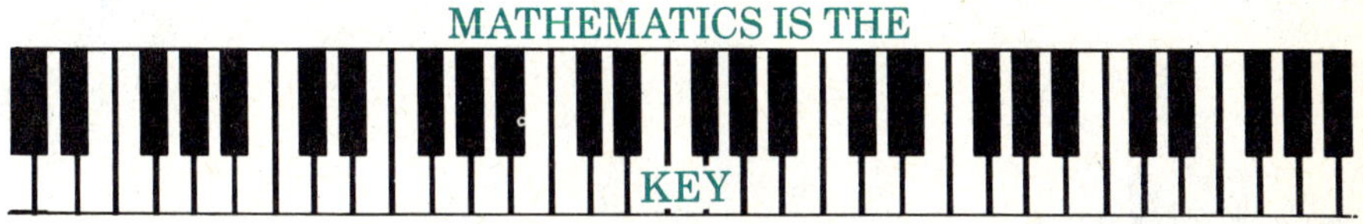

TO SUCCESS.

MILESTONE

1

BAND FORMATION

MANAGEMENT PLAN

On this ________ day of ___________________ in the year __________, I, ______________________________ plan to manage the band now known as _______________________ ________________________. The band has ______ members. I, the band manager, will begin by investing $41,000 and will make decisions regarding concerts, publicity, albums, and tours.

The band will pay a percentage of the band's net profit (to be agreed upon at a future date) as payment for the band manager's services.

ACTIVITY 1

READING AND WRITING WHOLE NUMBERS AND DECIMALS

Foundations: As a part of the rock music industry, you quickly learn to "think big." You will use large numbers while keeping track of concert attendance, record sales, and equipment expenses. This activity will give you practice in reading and writing numbers.

Directions: Statements 1–5 at the right contain numerals stated in words. Write the underlined words as *numerals* in the blanks provided.

♪ : Remember to insert commas and decimals correctly.

Statements 6–10 contain underlined numerals. Write each numeral in *words*.

Manager______________________ Date______________

1. Your first album sold one million sixty-one thousand fifty-two copies.

   ____

2. Fifty-seven thousand twenty-nine fans packed your last New York City concert.

   ____

3. A brand new deluxe electronic grand piano just hit the market. The list price is five thousand one hundred six dollars and twenty-four cents.

   ____

4. Last year's best-selling song was played on the radio nationally three hundred three million six thousand four hundred twelve times.

   ____

5. The E string on an electric guitar is twelve thousandths of an inch thick.

   ____

6. A good sound system costs $14,602.78 at a famous music store in San Francisco.

   ____

   ____

7. Last year your band traveled a total distance of 12,207 miles.

   ____

   ____

8. Your band's goal for this year is to sell 100,000,000 records.

   ____

   ____

9. The middle-C string on a grand piano is .062 of an inch thick.

   ____

   ____

10. Your band earned $6,752,008.03 from your last video.

    ____

    ____

Manager________________________ Date________________________

ACTIVITY 2
WRITING CHECKS

Foundations: The band manager pays all band expenses. In this activity, you will practice writing checks.

Directions: Complete the checks for each situation.

♪ : Remember to use the first check as an example.

1. On September 8, you booked Carnegie Hall for three New York City Concerts. The total booking fee was $12,053. Write a check payable to Carnegie Hall.

No. 2380

September 8 19-- 01-1234/0210

PAY TO THE ORDER OF Carnegie Hall $ 12,053.00

Twelve thousand fifty-three and 00/100 ——— Dollars

For Classroom Use Only

FIRST STATE BANK
10 STATE STREET
NEW YORK, NEW YORK 14614-1100

FOR Concert booking   Eve Watson

:021012349: 635 000 127

2. You want to start a new band. The three people you are considering as possible members live in San Francisco. If you pay their airfare, they will come to New York to discuss forming the band. Write a check payable to Southwestern Airlines for $1,375.45. Use today's date.

No. 2381

____________________ 19____ 01-1234/0210

PAY TO THE ORDER OF ______________________________ $ __________

____ Dollars

For Classroom Use Only

FIRST STATE BANK
10 STATE STREET
NEW YORK, NEW YORK 14614-1100

FOR ____________________   ____________________

:021012349: 635 000 127

3. A big rock star sued the last band you managed. He said your band's hit record sounded just like his. You went to court, lost the case, and now you must pay a fine. Write a check payable to Planet Records for $205,620. Use today's date.

No. 2382

_______________ 19____ 01-1234/0210

PAY TO THE ORDER OF _______________ $ _______

_______________ Dollars

For Classroom Use Only

FIRST STATE BANK
10 STATE STREET
NEW YORK, NEW YORK 14614-1100

FOR _______________ _______________

⑆0210123490⑆ 635 000 127⑈

4. You are buying an insurance policy to protect your musical instruments in case they are lost, destroyed, or stolen. Write a check to Felicidad Torres Insurance Agency for $6,083.44 as payment of one year's insurance. Use today's date.

No. 2383

_______________ 19____ 01-1234/0210

PAY TO THE ORDER OF _______________ $ _______

_______________ Dollars

For Classroom Use Only

FIRST STATE BANK
10 STATE STREET
NEW YORK, NEW YORK 14614-1100

FOR _______________ _______________

⑆0210123490⑆ 635 000 127⑈

Manager____________________ Date________________

ACTIVITY 3

ROUNDING WHOLE NUMBERS AND DECIMALS

Foundations: As band manager you will often be dealing with money calculations (such as percents and taxes). Because calculations often result in more decimal places than needed, rounding may be necessary. On busy bill-paying days, an electronic calculator will be a time-saver.

Directions: To the right are ten calculator display windows. Follow the instructions next to each display. Place your answers in the blanks provided at the right.

♪: Be sure to include a dollar sign since the calculator does not display one.

1. 147.23 Round to the nearest dollar. 1. __________
2. 1247.51 Round to the nearest dollar. 2. __________
3. 17.6289 Round to the nearest penny. 3. __________
4. 371.1271 Round to the nearest penny. 4. __________
5. 777.61 Round to the nearest dollar. 5. __________
6. 140.099 Round to the nearest penny. 6. __________
7. 139.90 Round to the nearest dollar. 7. __________
8. 7.7623 Round to the nearest penny. 8. __________
9. 1741.71 Round to the nearest dollar. 9. __________
10. 606.27 Round to the nearest dollar. 10. __________

ACTIVITY 4

ORDERING DECIMALS

Foundations: Strings must be placed on stringed instruments in ascending or descending order. The differences in string thicknesses are usually too small to see. The roadie must rely on reading the packages and ordering decimals.

Directions: Arrange the string sets for each of the instruments to the right in the correct order. Write the numbers on the blanks below the strings.

♪ : Remember, you are NOT changing the value of a decimal when you add zeroes at the right!

Manager____________________ Date________________

1. Arrange the string set for a bass guitar in ascending order:

   .1 .061 .08 .076

2. Arrange the string set for a rhythm guitar in descending order:

   .052 .02 .024 .043 .037 .017

3. Arrange the string set for a lead guitar in ascending order:

   .009 .046 .013 .024 .035 .02

4. Arrange the string set for an acoustic guitar in descending order:

   .043 .03 .011 .05 .021 .04

5. Arrange the string set for a five string bass guitar in ascending order:

   .08 .11 .071 .063 .055

6. Arrange the string set for a twelve string guitar in descending order:

   .049 .009 .037 .017 .015 .039 .04 .01
   .03 .024 .029 .052

Manager________________________ Date________________

ACTIVITY 5

DEPOSITS AND PAYMENTS

Foundations: You will need to know how to add and subtract decimals in order to make deposits and payments in your Check Register.

Directions: Complete the exercises at the right.

♪ : Remember to add deposits and subtract payments.

1. Write P for payment and D for deposit in the blank to the right of each item.

   a. Your band needed new sound equipment. You made the following purchases: $456.09, $238.78, and $23.85. 1a. ____________

   b. You owe the owner of a club $230 for failing to show up on time. b. ____________

   c. You receive a refund from Precision Drum Company because their salesperson overcharged you on your drum set. c. ____________

   d. You get your first paycheck from a club. d. ____________

   e. You pay your agent $300 for booking you at a big party. e. ____________

   f. You sell your grand piano for $3,000. f. ____________

   g. You buy a new bass amplifier for $456.98. g. ____________

2. During a very busy week in September, your band played at five clubs. The amount they made at each club is listed below.

   Sunday afternoon $120.00
   Sunday evening $340.00
   Tuesday night $180.00
   Friday night $489.78
   Saturday night $610.00

   a. On which *day* was the most money made? 2a. ____________

   b. On which *day* was the least money made? b. ____________

   c. What is the dollar amount difference between the lowest paying *performance* and the highest paying *performance?* c. ____________

3. On September 28, you had $3,178 in your Check Register. After the following transactions, what is your new balance?

   a. September 29 – $457.98 deposit 3a. ____________

   b. October 4 – $1,309.67 deposit b. ____________

c. October 10 – $909.97 payment — c. ______________

d. October 23 – four payments: $45.99, $234, $184.93, $284.11 — d. ______________

Manager____ Date____

CHECK REGISTER

Check Number	Date	Activity	Deposit	Payment	Balance	Teacher Check	Quality Points
		Start-up	$41,000.00		$41,000.00		
		Truck					
		Truck Options					
		Drums					
		Equipment					
		P.A. System					
		Lighting and Special Effects					

Manager____________________ Date____________________

CHECK REGISTER

Check Number	Date	Activity	Deposit	Payment	Balance	Teacher Check	Quality Points

Manager____________________ Date____________________

Check Register

Check Number	Date	Activity	Deposit	Payment	Balance	Teacher Check	Quality Points

Manager________________________________ Date________________________________

CHECK REGISTER

Check Number	Date	Activity	Deposit	Payment	Balance	Teacher Check	Quality Points

Manager________________________________ Date________________________

ACTIVITY 6

MULTIPLICATION OF WHOLE NUMBERS AND DECIMALS

Foundations: When your band travels, you will buy gas for the truck. You must be able to check the cost each time you get gas. The total cost which shows on the pump may not always be correct!

Directions: To the right are six gas pumps. The price of one gallon of gas is given above each pump. The number of gallons you bought shows on the front of each pump. On each gas pump, fill in the correct cost of the gas purchase in the boxes marked "Total Cost."

♪ : Be careful when placing the decimal point in your answer.

Manager________________________ Date________________

ACTIVITY 7

DIVISION WITH WHOLE NUMBERS AND DECIMALS

Foundations: Division is important in the music business. Whether you are behind the scenes or on the stage, you need to know how to divide.

Directions: Read each of the items to the right and write your answers in the blanks provided.

1. A new band is forming with eight members from the same high school. The members are chipping in equal amounts of money for a new sound system. The system costs $4,130. How much money does each member have to contribute?

   1. ________________

2. A certain musical group has three lead guitarists. Forty-two of their songs contain lead solos. If each guitarist must perform the same number of solos, how many solos will each guitarist perform?

   2. ________________

3. The Tempo Guitar Co. has produced 56,728 special electric guitars. Equal shipments of these guitars will be sent to 56 music stores. How many guitars will each store receive?

   3. ________________

4. A seven-member band made $14,084.91 selling rock T-shirts. If this money is divided equally, how much money does each member receive?

   4. ________________

5. The bill for three identical electric pianos totals $10,684.32. What is the price of each piano?

   5. ________________

6. A box of two dozen drumsticks costs $74.64. How much is one drumstick?

   6. ________________

7. Guitar pics can be sold by the gross (a gross equals 144 units). Each gross costs $17.28. How much does one pic cost?

   7. ________________

8. You are sending four identical amplifiers to Cincinnati, Ohio, to be repaired. You must pay the shipping costs. Shipping costs are based on the weight of the items being sent. If the total weight of the four amplifiers is 262 pounds, how much does each amplifier weigh?

   8. ________________

9. Vinny's Travel Agency has sent your band a bill for $2,178 to cover the cost of six round-trip tickets from New York to Los Angeles. How much does each round-trip ticket cost?

   9. ________________

Manager______________________________ Date____________________

ACTIVITY 8

ORDERING FRACTIONS

Foundations: The success of a band frequently depends on how often its songs reach the "Top 40." It is difficult to compare the success of different groups if they all haven't released the same number of songs. Fractions and decimals help us make these comparisons easily.

Directions: Fill in the missing entries in the chart at the right. The first row has been completed for you.

Group Name	Number of "Top 40" Hits	Number of Songs Released	$\frac{\text{Hits}}{\text{Songs Released}}$	Equivalent Decimal
The Beeps	10	40	$\frac{10}{40}$	.25
The Gross National Product	30	50		
The Long Island Sound	6	16		
The Moldy Oldies	56	64		
Route-80	5	8		
Practical Piggy & the Punk Pistols	16	32		

Directions: In the blanks to the right, list the bands in order from most successful to least successful according to the decimals in the last column of the above table.

♪ : Use your "zeroing-in" skills to list the bands in order.

First Place ______________________________

Second Place ______________________________

Third Place ______________________________

Fourth Place ______________________________

Fifth Place ______________________________

Sixth Place ______________________________

Manager________________________ Date____________________

ACTIVITY 9

UNDERSTANDING PERCENTS

Foundations: Percents are used in reporting information about bands, concerts, records, and many other music-related areas. As a band manager, you will often rely on your knowledge of percents.

Directions: Answer each of the questions at the right.

♪ : Remember expressions such as 5 out of 6 can be rewritten as $\frac{5}{6}$.

1. Four out of five members in your group never had music lessons. What percent of your group taught themselves how to play?

   1. ________________

2. Your band played to a standing room only (SRO) crowd at nine out of your last ten concerts. What percent of your concerts were SROs?

   2. ________________

3. Ten out of the 500 seats at a concert are called "obstructed view seats." These seats usually cost less money because parts of the stage cannot be seen when sitting there. What percent of the seats are obstructed view?

   3. ________________

4. During one of your shows, one of the five strings on your five-string bass guitar snapped.

   a. What percent of the original 5 strings broke? 4a. ________________

   b. What percent of the original 5 strings did not break? b. ________________

5. Fifteen out of your last fifty songs were written by the same person. What percent of your songs had the same author?

   5. ________________

6. Your band's latest album has been quite a success. Only two out of the eight songs on Side Two never made it to the "Top 40." What percent of the songs on Side Two never made it to the "Top 40"?

   6. ________________

7. Your sound engineer always brings 20 microphones along to each concert. Seventeen of those microphones are used to amplify the music. The rest are used for voices. What percent of the microphones are used for voices?

   7. ________________

8. If your band was signing a contract with a record company, you would want to know what percent of the company's songs became hits. Complete the following chart, and find out which company is the most successful. What is the most successful company?

   8. ________________

RECORD COMPANY	NO. OF HIT ALBUMS	NO. OF ALBUMS RELEASED	NO. OF HITS / NO. OF ALBUMS RELEASED	DECIMAL	PERCENT
Allison Records	6	25			
Penguin Records	14	56			
Braxton Records	26	100			
Waley Records	14	50			

ACTIVITY 10
FINDING PERCENTS

Foundations: Many situations in the music industry require you to know how to use percents. In your work this year, you will find the percent of many numbers.

Directions: Answer the questions to the right in the blanks provided.

Manager____________________ Date________________

1. The Jazz-King Guitar Company produced 1,250 guitars last year. Ten percent of the guitars had to be returned for repairs. How many of the guitars were returned?

   1. ________________

2. Your band played at a concert hall which seats 15,000 people. Ninety percent of the tickets were sold before the performance. How many tickets were sold before the performance?

   2. ________________

3. Your band made $12,120 from their last concert. As band manager, you received 12 percent of the profit. What was your share?

   3. ________________

4. Five percent of the tickets sold for your last tour were bought from a computerized ticket agency. If 125,000 tickets were sold, how many of them were computerized?

   4. ________________

5. The booking fee for the concert hall in Atlantic City, New Jersey, was $8,500. The security guards charged 7 percent of the booking fee. How much money was used to pay the guards?

   5. ________________

6. Twenty-five percent of the 800 people attending your last small concert were between the ages of 20 and 25. How many people were in that age group?

   6. ________________

7. Last year, you were forced to cancel a winter concert in Denver, Colorado, due to a heavy snowstorm. Fifteen thousand tickets had already been sold. Two percent of the ticket holders never returned for a refund. How many ticket holders never got their money back?

   7. ________________

ACTIVITY 11

CHANGING MIXED NUMBER PERCENTS TO EQUIVALENT DECIMALS

Manager____________________ Date________________

Foundations: The balance of money shown in your Check Register should be put in the bank that has the best interest rate. Banks often use mixed number percents to compute interest. It's your job to find the best interest rate.

Directions: Complete the equivalent decimal column in the chart at the right.

♪ : Remember to remove the percent sign (%) when writing the decimal.

BANK	INTEREST RATE PERCENT	EQUIVALENT DECIMAL
Shannon National Bank *The Green Machine*	$8\frac{1}{4}\%$	
Lewis Savings and Loan *You'll be glad you knew us*	$11\frac{3}{4}\%$	
The Snow Bank *Drift in for savings*	$5\frac{1}{2}\%$	
The Pi Bank *You're the apple of our eye*	$10\frac{1}{8}\%$	
First Foundation Bank *A sound investment*	$6\frac{5}{8}\%$	
Southwestern Savings *We want your business*	$12\frac{1}{2}\%$	
The Piggy Bank *Old-fashioned service*	$15\frac{1}{5}\%$	
The Burr Bank *We want your cold cash*	$9\frac{7}{8}\%$	
Data Bank *We compute your savings*	$7\frac{2}{5}\%$	

Manager____________________ Date____________________

ACTIVITY 12

PERCENTAGE DISCOUNT AND SALES TAX

Foundations: You are responsible for all equipment purchases, so you must be able to find discounts and calculate sales tax.

Directions: Determine the discount, sale price, sales tax, and total cost for each of the instruments at the right.

1. GUITAR
Originally $700

a. 11% discount — 1a. ________

b. Sale price — b. ________

c. 4% sales tax — c. ________

d. Total cost — d. ________

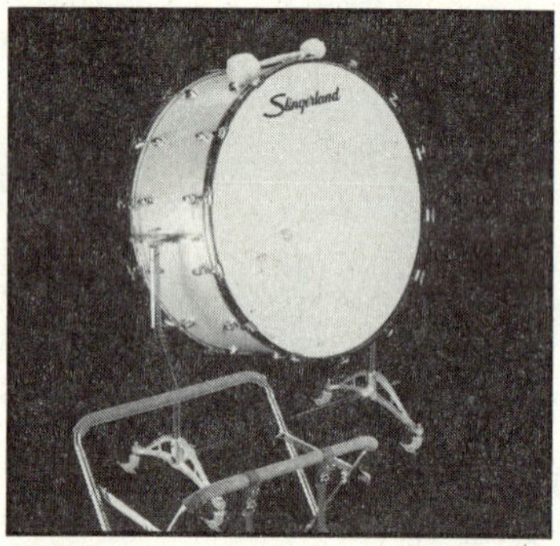

2. BASS DRUM
Originally $620

a. 35% discount — 2a. ________

b. Sale price — b. ________

c. 6% sales tax — c. ________

d. Total cost — d. ________

3. CYMBAL SET
Originally $730

a. 24% discount — 3a. ________

b. Sale price — b. ________

c. $6\frac{1}{2}$% sales tax — c. ________

d. Total cost — d. ________

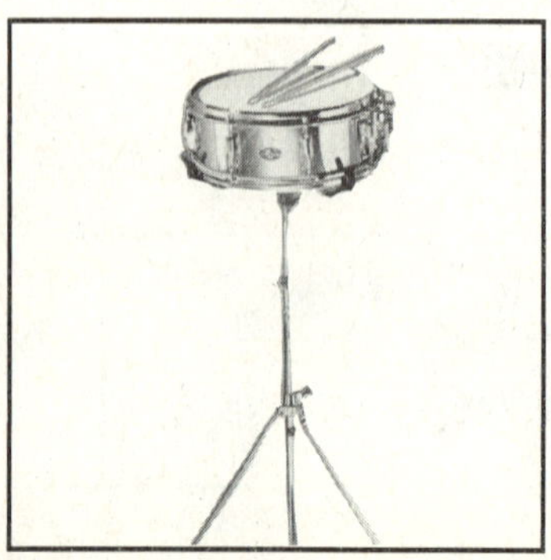

4. SNARE DRUM
Originally $78

a. 10% discount — 4a. ________

b. Sale price — b. ________

c. 8% sales tax — c. ________

d. Total cost — d. ________

Directions: Use the price list at the right to answer questions 5 through 9.

PRICE LIST

COMPLETE DRUM OUTFIT

ITEM	AMOUNT
Two 14″ × 24″ Bass Drums	$410
One 8″ × 12″ Tom-Tom	120
One 10″ × 14″ Tom-Tom	135
One 7″ × 14″ Chrome Snare Drum	147
One Pair of Bongos with Stand	133
Two Heavy-Duty Cymbal Stands	55
One Hi-Hat Cymbal	85
Two Cymbals	45

5. What is the regular cost of the complete drum outfit?

5. ________________

6. If the drum outfit is discounted 12%, how much would you save?

6. ________________

7. What is the sale price?

7. ________________

8. What would 6% sales tax be on the sale price?

8. ________________

9. What is the total price you would pay for the complete drum outfit?

9. ________________

Manager______________________________ Date______________________

ACTIVITY 13

READING CIRCLE GRAPHS

Foundations: In SOUND FOUNDATIONS, you must organize information in a clear manner. A circle graph is one way to present information clearly.

Part 1 Directions: Use the list at the right to answer questions 1 through 7.

	STADIUM	LOCATION	CAPACITY
1.	Arlington Stadium	Texas	35,698
2.	Astrodome	Texas	45,000
3.	Atlantic City Auditorium	New Jersey	40,000
4.	Atlanta-Fulton Stadium	Georgia	51,556
5.	Bismark Coliseum	North Dakota	7,000
6.	Busch Stadium	Missouri	50,100
7.	Candlestick Park	California	58,000
8.	Cow Palace	California	14,500
9.	Denver Coliseum	Colorado	9,038
10.	Dodger Stadium	California	56,000
11.	Exhibition Stadium	Toronto, Canada	40,001
12.	Fairgrounds Coliseum	Indiana	9,479
13.	Fenway Park	Massachusetts	33,465
14.	Halifax Metro Center	Nova Scotia, Canada	9,631
15.	Hershey Park Arena	Pennsylvania	7,286
16.	International Amphitheatre	Illinois	9,000
17.	Jacksonville Coliseum	Florida	7,900
18.	Kemper Memorial Arena	Missouri	16,320
19.	Kingdome	Washington	59,059
20.	Las Vegas Convention Center	Nevada	9,000
21.	Los Angeles Forum	California	17,505
22.	Louisiana Superdome	Louisiana	26,318
23.	Madison Square Garden	New York City	19,694
24.	Memorial Stadium	Maryland	52,137
25.	Milwaukee Arena	Wisconsin	10,934
26.	Municipal Stadium	Ohio	76,713
27.	Nassau Coliseum	New York	15,567
28.	Norfolk Scope	Virginia	10,600
29.	Oakland Coliseum	California	49,649
30.	Olympic Stadium	Montreal, Canada	58,838
31.	Omni	Georgia	15,389
32.	Pacific Coliseum	British Columbia, Canada	15,548
33.	Providence Civic Center	Rhode Island	11,619
34.	Quebec Coliseum	Quebec, Canada	10,531
35.	Reynolds Coliseum	North Carolina	11,431
36.	Richmond Coliseum	Virginia	10,700
37.	Riverfront Stadium	Ohio	51,963
38.	Royals Stadium	Missouri	40,635
39.	Salt Palace	Utah	12,201
40.	Sam Houston Coliseum	Texas	8,925
41.	San Diego Stadium	California	47,491
42.	Shea Stadium	New York	55,300
43.	Spectrum	Pennsylvania	18,276
44.	Three Rivers Stadium	Pennsylvania	50,235
45.	Tiger Stadium	Michigan	52,687
46.	Veteran's Stadium	Pennsylvania	58,651
47.	Veteran's Memorial Coliseum	Arizona	13,214
48.	Winnepeg Arena	Winnepeg, Canada	10,001
49.	Wrigley Field	Illinois	32,272
50.	Yankee Stadium	New York	54,028

1. How many stadiums have a capacity between 10,000 and 19,999?

1. ____________________

2. How many stadiums have a capacity of over 50,000?

2. ____________________

3. True/False: More than half of the stadiums have a capacity under 20,000.

3. ____________________

4. True/False: More than half of the stadiums have a capacity under 30,000.

4. ____________________

5. How many stadiums have a capacity of under 10,000?

5. ____________________

6. What *fraction* of the stadiums have a capacity of under 10,000?

6. ____________________

7. What *percent* of the stadiums have a capacity of under 10,000?

7. ____________________

Part 2 Directions: Use the graph to the right to answer the questions that are below it.

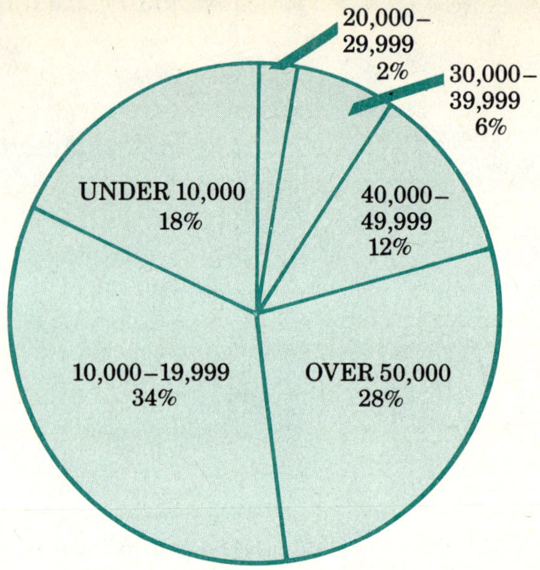

1. How many stadiums have a capacity between 10,000 and 19,999?

   1. ________________

2. How many stadiums have a capacity of over 50,000?

   2. ________________

3. True/False: More than half of the stadiums have a capacity under 20,000.

   3. ________________

4. True/False: More than half of the stadiums have a capacity under 30,000.

   4. ________________

5. How many stadiums have a capacity of under 10,000?

   5. ________________

6. What *fraction* of the stadiums have a capacity of under 10,000?

   6. ________________

7. What *percent* of the stadiums have a capacity of under 10,000?

   7. ________________

Manager______________________________ Date______________________

ACTIVITY 14

USING A PROTRACTOR

Foundations: As band manager, you must interpret information that is presented in music industry magazines. Circle graphs are often used to show large quantities of information in an easy-to-read form.

Directions: Use your protractor and the circle graph at the right to complete the chart below it.

♪: Remember the central angle of a circle has 360°.

EQUIPMENT BUDGET

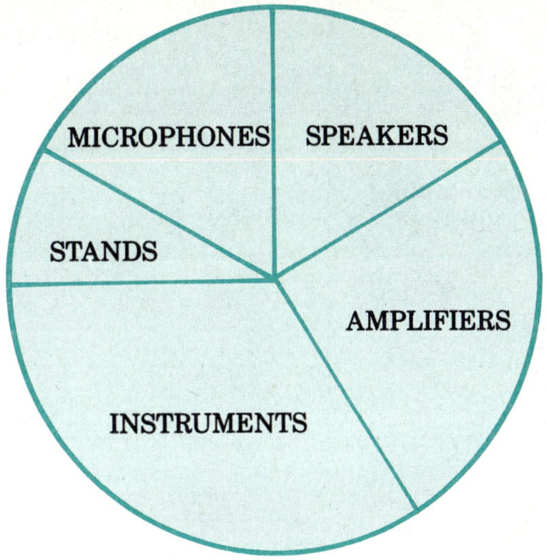

EQUIPMENT	NUMBER OF DEGREES	FRACTION	DECIMAL	PERCENT
Amplifiers				
Speakers				
Microphones				
Stands				
Instruments				

Manager_________________________ Date_________________

ACTIVITY 15

CONSTRUCTING A CIRCLE GRAPH SPINNER

Foundations: Managing a band requires many mathematical skills. However, a little bit of luck never hurt any band. You are now going to construct a circle graph spinner. Your band's luck on the dance club circuit is dependent on the luck of the spin.

Directions: The table to the right includes a list of dance clubs and percents. For each of the clubs listed, use the given percent to determine the number of degrees in a circle graph. Write the number of degrees for each in the "DEGREES" column. Then, using this information and your protractor, mark the sections and degrees on the circle on the next page to make a circle graph spinner.

♪: Remember there are 360° in the central angle of a circle.

DANCE CLUB	PERCENT	DEGREES
THE BROOKLYN BRIDGE	25%	
CAPRICORN'S	15%	
THE CITY ZOO	5%	
GLITTER'S	5%	
OMNIPOP	10%	
SHARK'S	5%	
STARZ	20%	
THE THIRD RAIL	15%	

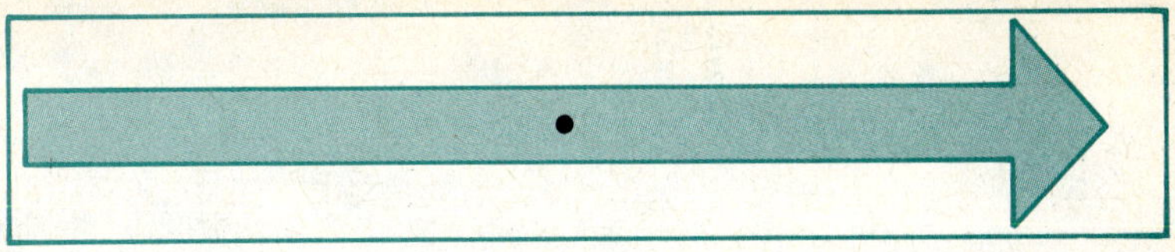

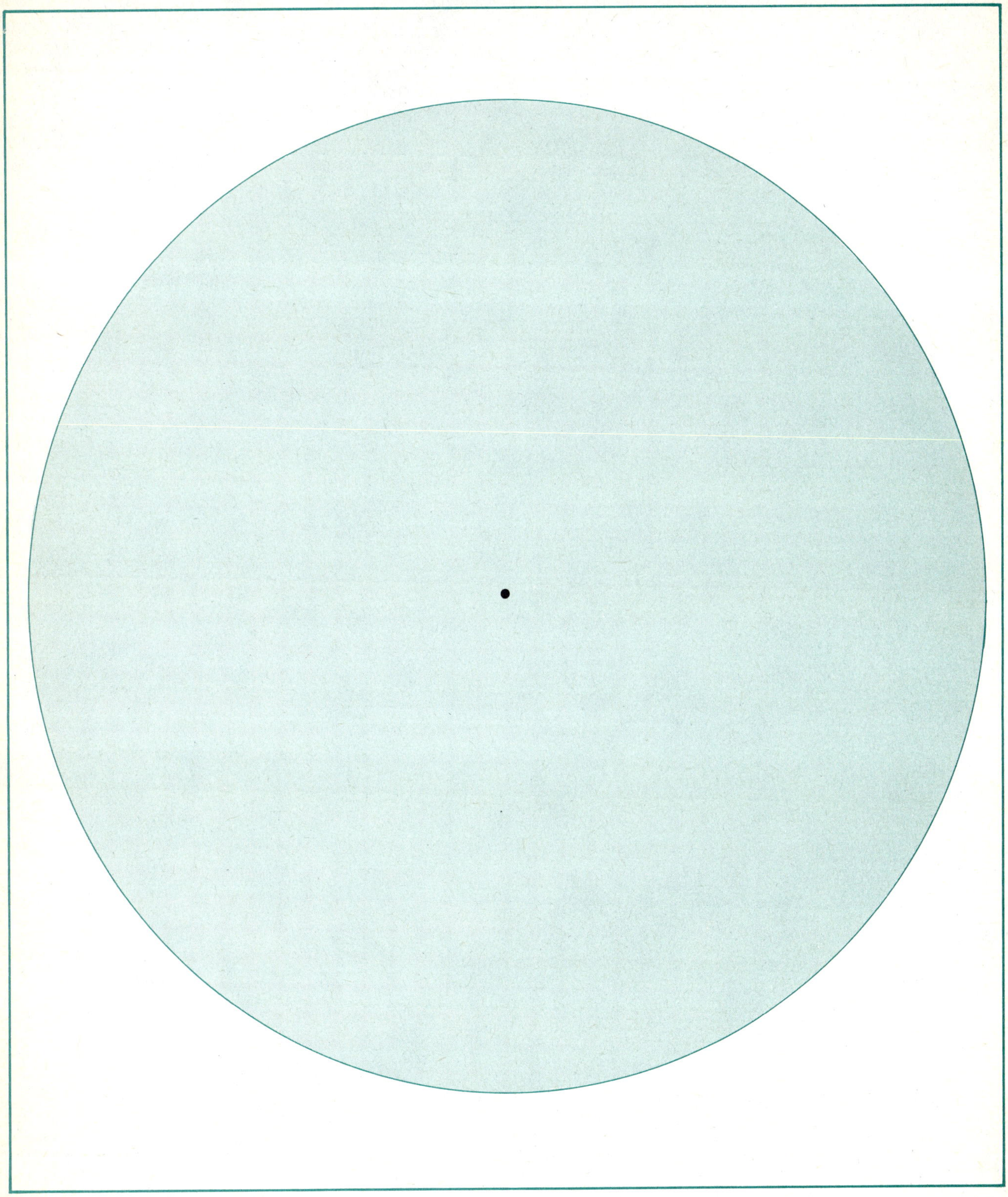

ACTIVITY 16

INTRODUCTION TO PROBABILITY

Foundations: When a well-known rock group comes to town for a concert, many contests are held. These contests may be sponsored by the local radio stations and concert halls. Probability will help you to understand your chances of winning these contests.

Directions: Answer the questions at the right.

Manager____________________ Date____________________

1. Radio station WEVE is giving away a set of concert tickets, and they have received 1,000 postcards. If you sent in one postcard, what is your chance of winning the tickets?

   1. ________________

2. A T-shirt promotion will be taking place at a sold-out New York City concert hall that seats 21,000 people. One seat will be selected at random, and that person will be awarded 20 T-shirts. What is the probability of your winning the T-shirts if you attend the concert?

   2. ________________

3. Radio WSFN is giving away concert tickets to one winner, and they have received 3,000 postcards. If you sent in 30 postcards, what is your chance of winning the ticket?

   3. ________________

4. Five hundred people have tickets for the next big concert in your town. One ticket holder will be selected at random to meet the band backstage. If you attend this concert, what are your chances of meeting the band?

   4. ________________

5. There are 500 balcony seats, 1,000 orchestra seats, and 300 mezzanine seats in a concert hall. You will be given a free ticket that is randomly selected.

   a. What is the probability of your getting an orchestra seat?

   5a. ________________

   b. What is the probability of your getting a balcony seat?

   b. ________________

   c. What is the probability of your getting a mezzanine seat?

   c. ________________

   d. What is the probability of your NOT getting an orchestra seat?

   d. ________________

   e. Each of the three sections (balcony, orchestra, mezzanine) has a front row. The balcony has 40 front row seats, the orchestra has 50 front row seats, and the mezzanine has 18 front row seats. What is the probability of your free ticket being a front row seat?

e. ________________

ACTIVITY 17

PROBABILITY AND PERCENTS

Manager____________________ Date________________

Foundations: The luck factor in certain activities in SOUND FOUNDATIONS will be simulated by spins of your circle graph spinner. The purpose of this activity is to see how probability is related to the circle graph sectors.

Part 1 Directions: Spin your circle graph spinner 100 times and record the results in the chart at the right.

♪ : When you get *near* 100 spins, check to make sure that you do not spin more than 100 times!

DANCE CLUB NAME	TALLEY
THE BROOKLYN BRIDGE	
CAPRICORN'S	
THE CITY ZOO	
GLITTER'S	
OMNIPOP	
SHARK'S	
STARZ	
THE THIRD RAIL	

Part 2 Directions: Complete the chart below with the results from the entire class. Follow the directions that your teacher will give you so that the chart will be accurate.

	Student's Name	THE BROOKLYN BRIDGE	CAPRICORN'S	THE CITY ZOO	GLITTER'S	OMNIPOP	SHARK'S	STARZ	THE THIRD RAIL	RUNNING TOTAL OF SPINS
1										100
2										200
3										300
4										400
5										500
6										600
7										700
8										800
9										900
10										1,000
11										1,100
12										1,200
13										1,300
14										1,400
15										1,500
16										1,600
17										1,700
18										1,800
19										1,900
20										2,000
21										2,100
22										2,200
23										2,300
24										2,400
25										2,500
26										2,600
27										2,700
28										2,800
29										2,900
30										3,000
31										3,100
32										3,200
33										3,300
34										3,400
TOTAL SPINS FOR EACH CLUB										
FRACTION OF TOTAL SPINS										
PERCENT										

Manager______________________________ Date______________________

ACTIVITY 18

MEAN, MEDIAN, AND MODE

Foundations: In the early days of your band, progress is best measured by the number of people you draw into your performances. Band managers must be able to keep records and check club attendance over any period of time. If more people come to see your band, your band can charge more money to perform.

Directions: To the right is a table with attendance figures for your band's last nine performances at Glitter's. Answer the questions below the table using the information in the table.

MONTH	TOTAL ATTENDANCE	EVENT
January	450	New Year's Eve Dance
February	122	President's Dance
March	271	Spring Dance
April	512	April Fool's Dance
May	450	Memorial Day Dance
June	311	Summer Dance
July	301	Independence Day Dance
August	450	Pool Party
September	311	Labor Day Dance

1. The manager of Glitter's may give your band a raise. The manager must figure the mean attendance. What is the mean attendance over the last nine months?

   1. ________________

2. What is the mean attendance over the last four months?

   2. ________________

3. Which time period would you want Glitter's to base your raise on, the last nine months or the last four months?

   3. ________________

4. What was the mean attendance during the winter months (January, February, March)?

   4. ________________

5. What is the median attendance for all nine months?

   5. ________________

6. The February President's Dance had poor attendance because of a blizzard.

   a. What is the median attendance for the eight remaining months (excluding February)? 6a. ________________

   b. What is the mean attendance for the eight remaining months (excluding February)? b. ________________

7. What is the mode attendance figure for the last nine months?

7. ____________________

8. Your band is scheduled to play Glitter's Halloween Costume Party in October. If the attendance that night is 400, will your mean attendance (from question 1) go up or down?

8. ____________________

MILESTONE 2

EQUIPMENT PURCHASE

ACTIVITY 19
TRUCK PURCHASE

Manager____________________ Date________________

Foundations: Your first purchase as a band manager is a truck. You will use the truck to transport your band's equipment.

Directions: Complete the Practice Version of Milestone 2/ Form A – Truck form.

♪ : Round to the nearest cent.

PRACTICE VERSION

MILESTONE 2
FORM A

TRUCK

STOCK NUMBER	DESCRIPTION	LIST PRICE
C3W4628	CHASSIS & CAB	$7,215.38
671W2T	HEAVY-DUTY FRONT STABILIZER BAR	41.91
192672	HEAVY-DUTY REAR SPRINGS	161.00
4STMW	4-SPEED MANUAL TRANSMISSION	N/C
2B7EP247	AUXILIARY FUEL TANK	192.00
P76EVE	POWER STEERING	351.77
Z-26W	DUAL REAR WHEELS	143.11
76214	AIR CONDITIONING	666.77
73112S	AM/FM STEREO WITH CASSETTE	320.19
T-101	DIGITAL TACHOMETER	77.12
7762	BODY-SIZE MOULDING	N/C
1161	GAUGES	45.00
	TOTAL LIST PRICE	
	11% DISCOUNT	
	SALE PRICE	
	7% SALES TAX	
	TOTAL TRUCK COST	

Manager_______________________ Date_______________________

MILESTONE 2
FORM A

TRUCK

STOCK NUMBER	DESCRIPTION	LIST PRICE
6104	CHASSIS & CAB	$7,129.74
7111	HEAVY-DUTY SHOCK ABSORBERS	101.00
6124	POWER STEERING	212.11
101H	TACHOMETER	77.64
11AT	AUTOMATIC TRANSMISSION	911.00
TW-01	TINTED WINDSHIELD	44.11
SPR7	HEAVY-DUTY SPRINGS	324.65
AR61	ALARM	425.00
7716	GAUGES	165.00
661S	EMISSION CONTROL	N/C
Z1UC	UNDERCOATING	212.00
BS11	VINYL BENCH SEAT	N/C
	TOTAL LIST PRICE	
	9% DISCOUNT	
	SALE PRICE	
	8% SALES TAX	
	TOTAL TRUCK COST	

Write a check payable to Watson Motors.

No. 2384

_______________________ 19____ 01-1234/0210

PAY TO THE ORDER OF _______________________ $ ________

_______________________ Dollars

For Classroom Use Only

FIRST STATE BANK
10 STATE STREET
NEW YORK, NEW YORK 14614-1100

FOR _______________ _______________

⑆0210123491⑆ 635 000 127⑈

SEE ACTIVITY 19 FOR PRACTICE VERSION

ACTIVITY 20
TRUCK OPTIONS

Manager________________________ Date________________

Foundations: You are now going to purchase extra equipment for the truck.

Directions: Use the catalog on pages 53 and 54 to complete the form below.

♪ : Pay careful attention to the quantity column.

PRACTICE VERSION

MILESTONE 2
FORM B

TRUCK OPTIONS

ITEM	CATALOG NUMBER	LIST PRICE	QUANTITY	TOTAL
Sofa Sleeper	2439-UREW		1	
Sunroof	2376-0054		1	
Map Light	334-W4T6		2	
Speaker	2746-9TTT		3	
Chrome Wheels	1037-3866		4	
Luggage Rack	2275-6644		1	
Gas Can	2759-46WW		1	
			TOTAL	
			15% DISCOUNT	
			SALE PRICE	
			8% SALES TAX	
			TOTAL COST	

Manager____ Date____

MILESTONE 2
FORM B

TRUCK OPTIONS

♪ : Remember to spend between $200 and $2,000 on one to seven options from the truck options catalog on pages 53 and 54.

ITEM	CATALOG NUMBER	LIST PRICE	QUANTITY	TOTAL

TOTAL	
10% DISCOUNT	
SALE PRICE	
8% SALES TAX	
TOTAL COST	

Write a check payable to Art & Sylvia's Truck Supplies.

No. 2385

______________________ 19____ 01-1234/0210

PAY TO THE ORDER OF ____ $ __________

____ Dollars

For Classroom Use Only

FIRST STATE BANK
10 STATE STREET
NEW YORK, NEW YORK 14614-1100

FOR ______________________ ______________________

⑆021012349⑆ 635 000 127⑈

SEE ACTIVITY 20 FOR PRACTICE VERSION

TRUCK OPTIONS CATALOG

SUN ROOF

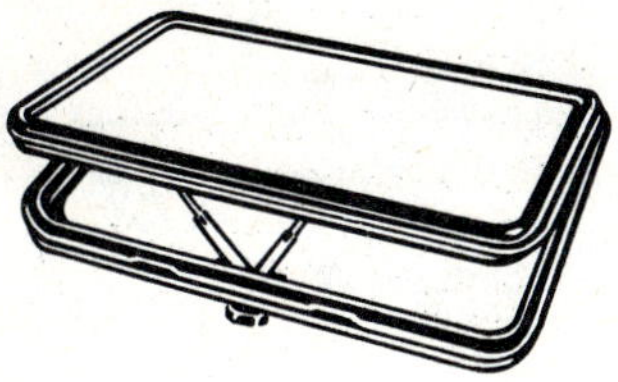

2376-0054 . . . . $176.60

SAE & METRIC SOCKET WRENCH SETS

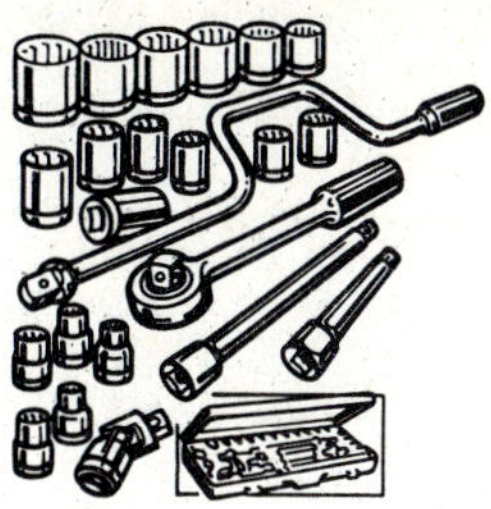

1983-4452 . . . . $17.95

DIGITAL CLOCK

7195-8377 . . . . $34.55

AM/FM MULTIPLEX STEREO RADIO

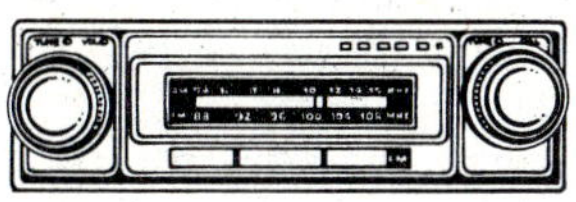

7465-392 . . . . $424.90

MAP LIGHT

334-W4T6 . . . . $15.00

GAS CAN

2759-46WW $16.98

LUGGAGE RACK

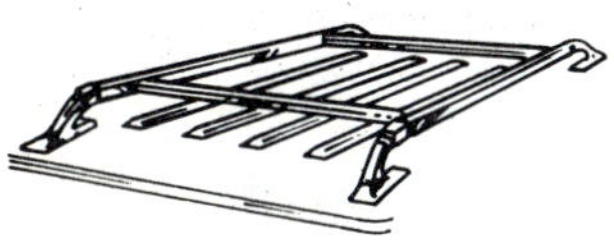

2275-6644 . . . . $85.00

CHROME WHEEL

1037-3866 . . . . $201.50

MAGNETIC SPEAKERS

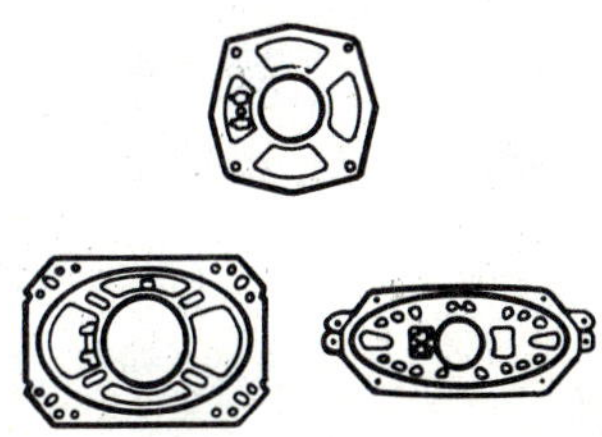

2746-9TTT $166.50

BUMPER GUARDS

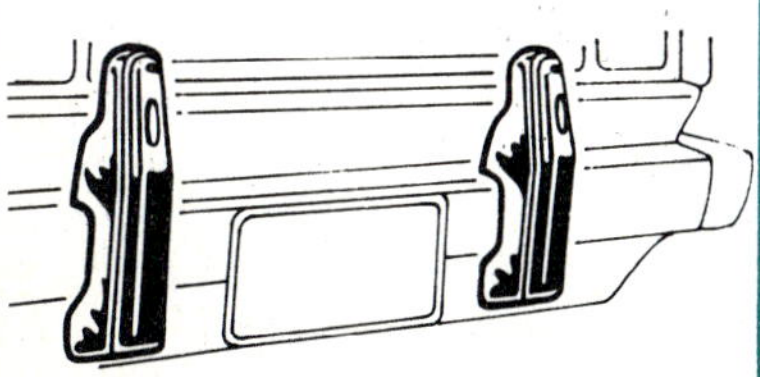

8847-5RSF . . . . . . $22.98 /pair

HYDRAULIC SERVICE JACK

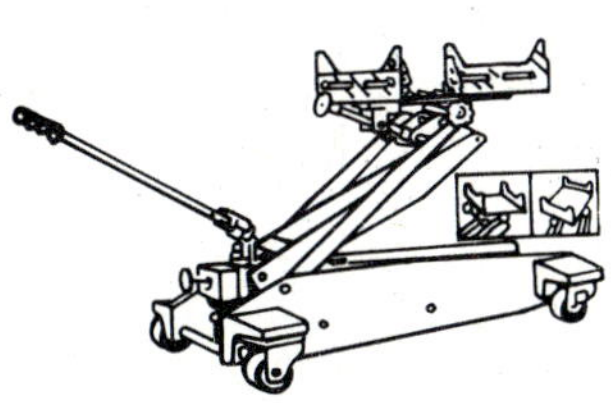

7295-33HJ $299.20

DUAL SLIDING SIDE WINDOWS

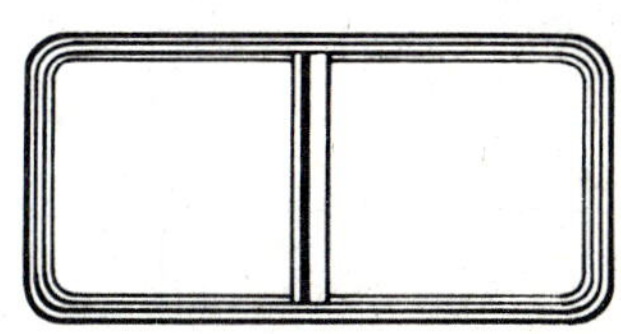

2750-SW55 . . . . $68.95

CARPETS

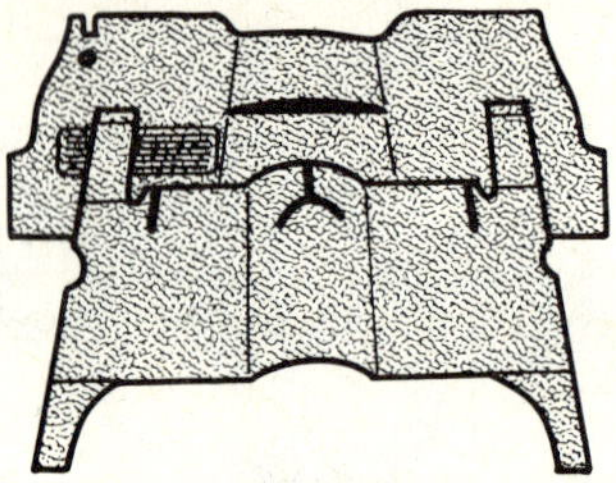

2345-7YYT $98.00/set

DELUXE STEREO POWER AMPLIFIER

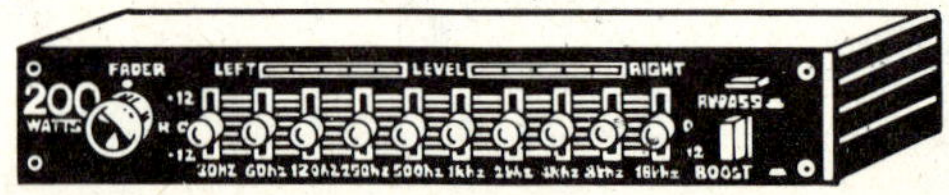

7766-ANY5 . . . . $59.95

POWER ANTENNA

1199-OHBB $24.99

STEERING WHEEL AND BRAKE LOCK

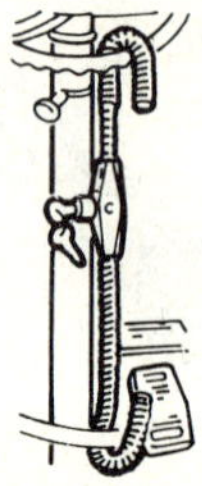

4433-8HGT $12.95

FLOOR MATS

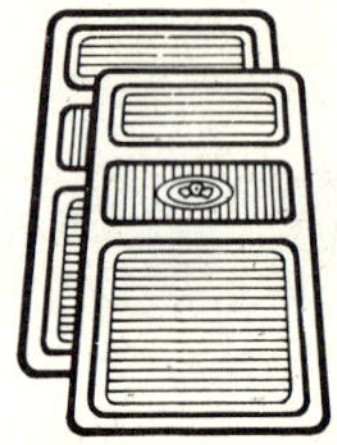

4243-TTV6 . . . . . $14.95/pair

SHEEPSKIN CAR SEAT COVERS

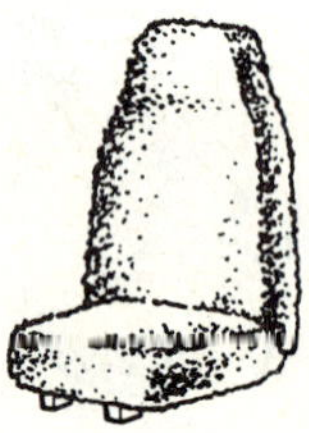

9453-7296 . . . . $150.00

40-CHANNEL CB MONITOR

2583-TWI8 . . . . $189.80

SOFA SLEEPER

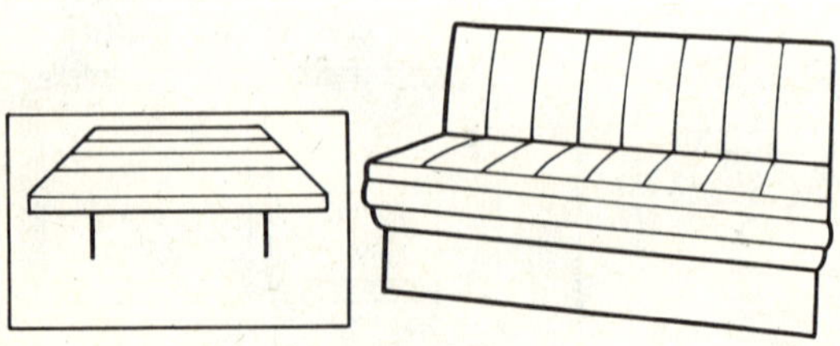

2439-UREW $499.95

STEERING WHEEL COVER

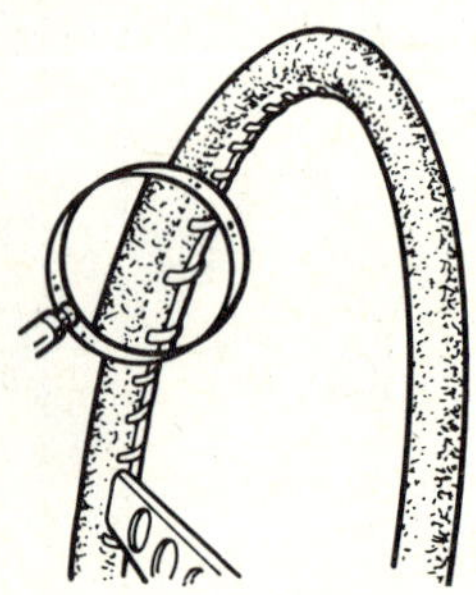

1856-3833 . . . . $12.00

Manager________________________________ Date________________________________

MILESTONE 2
FORM C

DRUMS

Two 14″ × 24″ Bass Drums
@ $200 . . . . $400.00
Telescopic Spurs, 2 pr. ea.
Bass Drum 40.00
8″ ×12″ Tom Tom 100.00
9″ × 13″ Tom Tom 104.00
10″ × 14″ Tom Tom 107.00
10″ × 15″ Tom Tom 110.00
Two Double Tom Holders 80.00
6½″ × 14″ Chrome Snare Drum 125.00
16″ × 16″ Tom Tom with Legs 145.00
18″ × 20″ Tom Tom with Legs 200.00
Hand Bongos with Stand 120.00
Two Anchors for Bass Drum @ $7 . . . 14.00
Two Super Pedals @ $50 . . . . 100.00
Two Heavy Duty Cymbal
Stands @ $32 . . . . 64.00
Hi Hat Stand 70.00
Snare Stand 35.00
Throne 44.00
Brushes, pair 3.00
Sticks, pair 3.00

TOTAL ____________________

12% DISCOUNT ____________________

SALE PRICE ____________________

6% SALES TAX ____________________

TOTAL COST ____________________

Write a check payable to
Leslie's Drum Center.

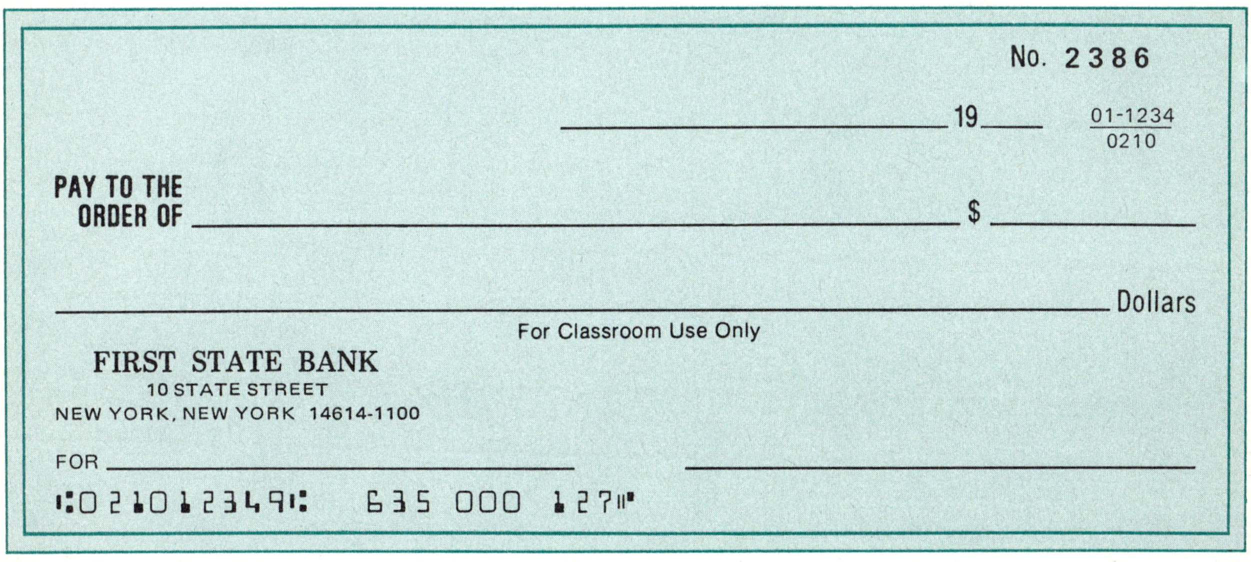

Manager____________________ Date____________________

MILESTONE 2
FORM D

EQUIPMENT

♪ : Remember to spend between $6,000 and $15,000.

From the order form below, you must choose AT LEAST

1 Bass guitar
1 Electric piano
3 Guitar amplifiers
2 Bass amplifiers
4 Guitars

ITEM	LIST PRICE	QUANTITY	TOTAL
BASS GUITARS			
Jazzman Bass Guitar	$ 465		
Rickyman Bass Guitar	$ 405		
Forrester Bass Guitar	$ 387		
Big Boss Bass Guitar	$ 410		
ELECTRIC PIANOS			
Lawson Electric Piano	$3,100		
Kay Baby Grand Piano	$1,100		
Planet Piano	$ 761		
GUITAR AMPLIFIERS			
Soundmaster Guitar Amplifier	$ 471		
Lawson Guitar Amplifier	$ 753		
Big Beep Guitar Amplifier	$ 621		
Planet Guitar Amplifier	$1,040		
BASS AMPLIFIERS			
Bass King Bass Amplifier	$ 437		
Soundmaster Bass Amplifier	$ 698		
Lawson Super Bass Amplifier	$1,131		
GUITARS			
Lawson Acoustic Guitar	$ 750		
Soundmaster Electric Guitar	$ 802		
Big Boss Electric Guitar	$ 950		
Stratosphere Guitar	$1,500		

TOTAL	
14% DISCOUNT	
SALE PRICE	
5% SALES TAX	
TOTAL COST	

Manager____________________ Date____________________

MILESTONE 2
FORM E

BIG P.A.

DESCRIPTION	CATALOG NUMBER	LIST PRICE	QUANTITY	TOTAL PRICE
Speaker Cabinets	RS101	$ 400.00	2	
Speaker Cabinets	RG306	$ 611.00	2	
Horns	BG421	$ 190.00	2	
Audio Console	LS101	$1,079.00	1	
Power Amplifier	NG107	$ 416.00	5	
Microphones	RKG-1972	$ 141.92	8	
Microphone Stands	1957-210	$ 32.50	8	
			TOTAL	
			13% DISCOUNT	
			SALE PRICE	
			8% SALES TAX	
			TOTAL COST	

Manager________________________ Date________________________

MILESTONE 2
FORM F

SMALL P.A.

DESCRIPTION	CATALOG NUMBER	LIST PRICE	QUANTITY	TOTAL PRICE
Monitor Columns	AS1977	$210.00	2	
Speaker System	ES1978	$623.00	2	
Audio Console	T42ME4U	$685.00	1	
Microphones	RKG-1957	$ 98.50	5	
Microphone Stands	1957-210	$ 32.50	5	
			TOTAL	
			15% DISCOUNT	
			SALE PRICE	
			8% SALES TAX	
			TOTAL COST	

Manager____ Date____________________________________

MILESTONE 2
FORM G

LIGHTING & SPECIAL EFFECTS

♪ : Remember to spend $1,200 or less on one to seven different items from the Lighting & Special Effects catalog on pages 65 and 66.

ITEM	LIST PRICE	QUANTITY	TOTAL PRICE
		TOTAL	
		14% DISCOUNT	
		SALE PRICE	
		$7\frac{1}{2}$% SALES TAX	
		TOTAL COST	

BORDERLIGHTS

BLST-700 . . . . $249.00

ELECTRIC SMOKE BOX

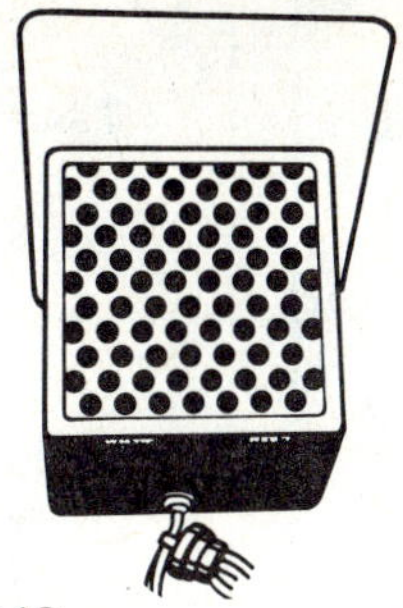

SEMN-243 . . . . $46.50

BUBBLE MACHINE

DNTA-266 . . . . $299.50

FOG MACHINE

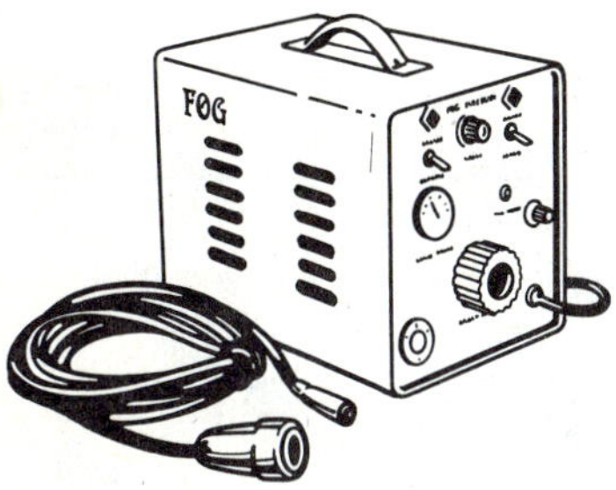

FGMST-822 . . . . $525.00

MINI-KIT

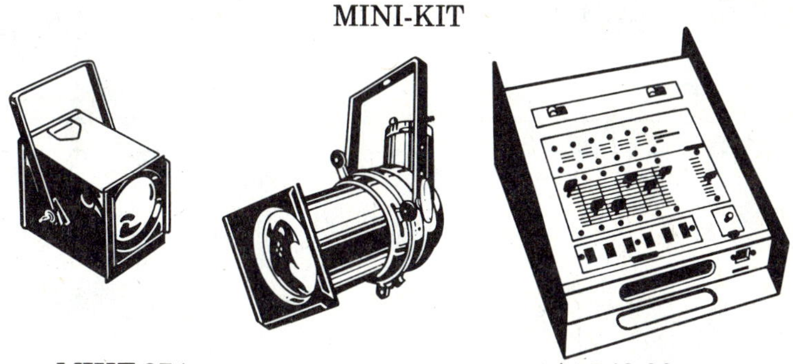

MIKT-274 . . . . $1,249.00

MIRRORBALL

MBMB-633 . . . . $249.50

SNOW MACHINE

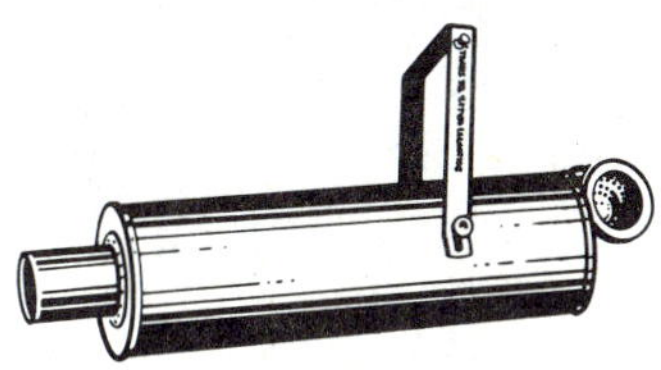

SNWM-388 . . . .$299.00

16″ BEAM PROJECTOR

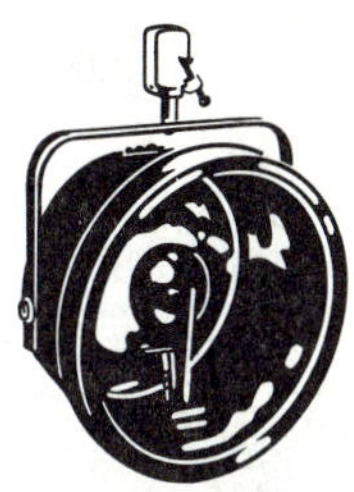

BMPR-444 . . . . $139.00

INTERCOM SYSTEM

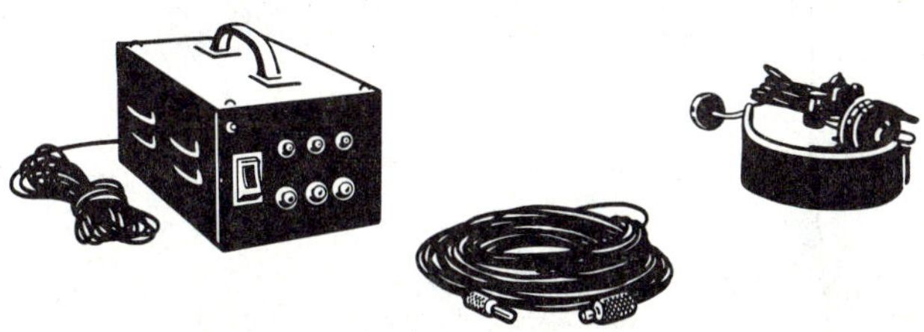

ICSS-999 . . . . $450.00

LIGHTING CONTROL BOARD

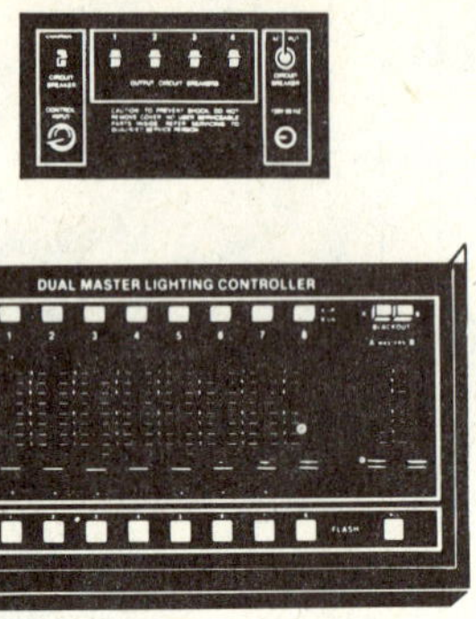

GWST-979 . . . . $249.95

LITE KIT

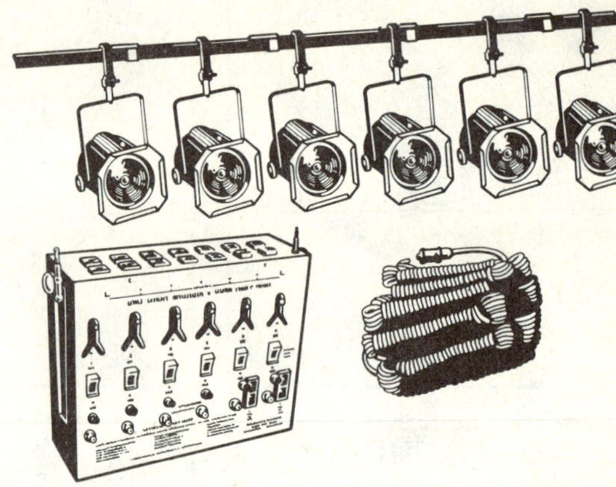

GWST-112 . . . . $946.00

SPOTLIGHT

SLTY-275 . . . . $509.50

STROBE LIGHT PACKAGE

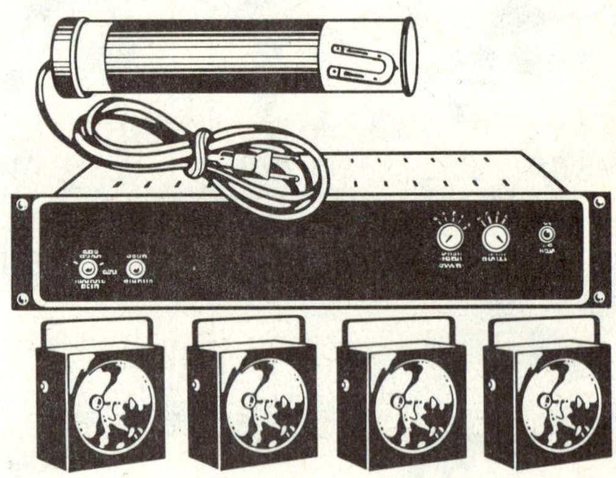

SSKJ-194 . . . . $453.99

SPIN BEAMS

SBNM-275 . . . . $285.00

COMPLETE SHOW PACKAGE

CSPK $759.99

MILESTONE 3

REHEARSAL

ACTIVITY 21

REHEARSAL FORMS

Foundations: Milestone 3/Form H – Rehearsal form has directions that are similar to the directions found on income tax forms. After successful completion of a Milestone 3/Form H – Rehearsal form, you will receive quality points.

Directions: Using this information, complete the following for Michelle Crum: Michelle Crum correctly answered four out of five questions on today's quiz. She has decided that her band will practice seven hours.

Manager____________________ Date______________

PRACTICE VERSION 1

MILESTONE 3
FORM H

REHEARSAL

♪ : You may rent a studio for one to ten hours.

1. How many problems did you answer correctly on the quiz that was returned to you today? 1. ____________

2. How many questions were asked on the quiz? 2. ____________

3. Represent the part of the quiz you answered correctly as a fraction. 3. ____________

4. Change the fraction in No. 3 to an equivalent decimal. 4. ____________

5. Round the decimal in No. 4 to the nearest hundredth. 5. ____________

6. Change the decimal in No. 5 to an equivalent percent. *This number represents the percentage of productive rehearsal time.* 6. ____________

7. How many hours will your band rehearse today? 7. ____________

8. If rehearsal studio time costs \$28.25 per hour, then what will today's rehearsal cost? 8. \[\]

9. Find the number of productive rehearsal hours by multiplying total hours rehearsed (No. 7) by the percentage of productive rehearsal time (No. 6). 9. ____________

10. Round the number of productive rehearsal hours (No. 9) to the nearest integer. *This number represents the number of productive rehearsal hours.* 10. ____________

11. Each *productive rehearsal hour* (No. 10) is worth two quality points. How many quality points do you receive for today's rehearsal? 11. ____________

Directions: Using this information, complete the following for Rick Chen: Rick Chen correctly answered three out of five questions on today's quiz, and he wants to rehearse for ten hours.

PRACTICE VERSION 2

MILESTONE 3
FORM H

REHEARSAL

♪ : You may rent a studio for one to ten hours.

1. How many problems did you answer correctly on the quiz that was returned to you today? 1. ______________
2. How many questions were asked on the quiz? 2. ______________
3. Represent the part of the quiz you answered correctly as a fraction. 3. ______________
4. Change the fraction in No. 3 to an equivalent decimal. 4. ______________
5. Round the decimal in No. 4 to the nearest hundredth. 5. ______________
6. Change the decimal in No. 5 to an equivalent percent. This number represents the percentage of productive rehearsal time. 6. ______________
7. How many hours will your band rehearse today? 7. ______________
8. If rehearsal studio time costs $28.25 per hour, then what will today's rehearsal cost? 8. ☐
9. Find the number of productive rehearsal hours by multiplying total hours rehearsed (No. 7) by the percentage of productive rehearsal time (No. 6). 9. ______________
10. Round the number of productive rehearsal hours (No. 9) to the nearest integer. This number represents the number of productive rehearsal hours. 10. ______________
11. Each *productive rehearsal hour* (No. 10) is worth two quality points. How many quality points do you receive for today's rehearsal? 11. ______________

Manager________________________________ Date________________________________

MILESTONE 3
FORM H

REHEARSAL

♪ : You may rent a studio for one to ten hours.

1. How many problems did you answer correctly on the quiz that was returned to you today? 1. ________

2. How many questions were asked on the quiz? 2. ________

3. Represent the part of the quiz you answered correctly as a fraction. 3. ________

4. Change the fraction in No. 3 to an equivalent decimal. 4. ________

5. Round the decimal in No. 4 to the nearest hundredth. 5. ________

6. Change the decimal in No. 5 to an equivalent percent. This number represents the percentage of productive rehearsal time. 6. ________

7. How many hours will your band rehearse today? 7. ________

8. If rehearsal studio time costs $28.25 per hour, then what will today's rehearsal cost? 8. ________

9. Find the number of productive rehearsal hours by multiplying total hours rehearsed (No. 7) by the percentage of productive rehearsal time (No. 6). 9. ________

10. Round the number of productive rehearsal hours (No. 9) to the nearest integer. This number represents the number of productive rehearsal hours. 10. ________

11. Each *productive rehearsal hour* (No. 10) is worth two quality points. How many quality points do you receive for today's rehearsal? 11. ________

12. Record the payment in your Check Register.

SEE ACTIVITY 21 FOR PRACTICE VERSION

Manager____________________ Date____________________

MILESTONE 3
FORM H

REHEARSAL

♪ : You may rent a studio for one to ten hours.

1. How many problems did you answer correctly on the quiz that was returned to you today? 1. __________

2. How many questions were asked on the quiz? 2. __________

3. Represent the part of the quiz you answered correctly as a fraction. 3. __________

4. Change the fraction in No. 3 to an equivalent decimal. 4. __________

5. Round the decimal in No. 4 to the nearest hundredth. 5. __________

6. Change the decimal in No. 5 to an equivalent percent. This number represents the percentage of productive rehearsal time. 6. __________

7. How many hours will your band rehearse today? 7. __________

8. If rehearsal studio time costs $28.25 per hour, then what will today's rehearsal cost? 8. __________

9. Find the number of productive rehearsal hours by multiplying total hours rehearsed (No. 7) by the percentage of productive rehearsal time (No. 6). 9. __________

10. Round the number of productive rehearsal hours (No. 9) to the nearest integer. This number represents the number of productive rehearsal hours. 10. __________

11. Each *productive rehearsal hour* (No. 10) is worth two quality points. How many quality points do you receive for today's rehearsal? 11. __________

12. Record the payment in your Check Register.

SEE ACTIVITY 21 FOR PRACTICE VERSION

Manager____________________ Date____________________

MILESTONE 3
FORM H

REHEARSAL

♪ : You may rent a studio for one to ten hours.

1. How many problems did you answer correctly on the quiz that was returned to you today? 1. ____________
2. How many questions were asked on the quiz? 2. ____________
3. Represent the part of the quiz you answered correctly as a fraction. 3. ____________
4. Change the fraction in No. 3 to an equivalent decimal. 4. ____________
5. Round the decimal in No. 4 to the nearest hundredth. 5. ____________
6. Change the decimal in No. 5 to an equivalent percent. This number represents the percentage of productive rehearsal time. 6. ____________
7. How many hours will your band rehearse today? 7. ____________
8. If rehearsal studio time costs $28.25 per hour, then what will today's rehearsal cost? 8. \[\]
9. Find the number of productive rehearsal hours by multiplying total hours rehearsed (No. 7) by the percentage of productive rehearsal time (No. 6). 9. ____________
10. Round the number of productive rehearsal hours (No. 9) to the nearest integer. This number represents the number of productive rehearsal hours. 10. ____________
11. Each *productive rehearsal hour* (No. 10) is worth two quality points. How many quality points do you receive for today's rehearsal? 11. ____________
12. Record the payment in your Check Register.

SEE ACTIVITY 21 FOR PRACTICE VERSION

Manager____________________ Date____________________

MILESTONE 3
FORM H

REHEARSAL

♪ : You may rent a studio for one to ten hours.

1. How many problems did you answer correctly on the quiz that was returned to you today? 1. __________
2. How many questions were asked on the quiz? 2. __________
3. Represent the part of the quiz you answered correctly as a fraction. 3. __________
4. Change the fraction in No. 3 to an equivalent decimal. 4. __________
5. Round the decimal in No. 4 to the nearest hundredth. 5. __________
6. Change the decimal in No. 5 to an equivalent percent. This number represents the percentage of productive rehearsal time. 6. __________
7. How many hours will your band rehearse today? 7. __________
8. If rehearsal studio time costs $28.25 per hour, then what will today's rehearsal cost? 8. [ ]
9. Find the number of productive rehearsal hours by multiplying total hours rehearsed (No. 7) by the percentage of productive rehearsal time (No. 6). 9. __________
10. Round the number of productive rehearsal hours (No. 9) to the nearest integer. This number represents the number of productive rehearsal hours. 10. __________
11. Each *productive rehearsal hour* (No. 10) is worth two quality points. How many quality points do you receive for today's rehearsal? 11. __________
12. Record the payment in your Check Register.

SEE ACTIVITY 21 FOR PRACTICE VERSION

Manager________________________________ Date________________________________

MILESTONE 3
FORM H

REHEARSAL

♪ : You may rent a studio for one to ten hours.

1. How many problems did you answer correctly on the quiz that was returned to you today? 1. ____________
2. How many questions were asked on the quiz? 2. ____________
3. Represent the part of the quiz you answered correctly as a fraction. 3. ____________
4. Change the fraction in No. 3 to an equivalent decimal. 4. ____________
5. Round the decimal in No. 4 to the nearest hundredth. 5. ____________
6. Change the decimal in No. 5 to an equivalent percent. This number represents the percentage of productive rehearsal time. 6. ____________
7. How many hours will your band rehearse today? 7. ____________
8. If rehearsal studio time costs $28.25 per hour, then what will today's rehearsal cost? 8. [ ]
9. Find the number of productive rehearsal hours by multiplying total hours rehearsed (No. 7) by the percentage of productive rehearsal time (No. 6). 9. ____________
10. Round the number of productive rehearsal hours (No. 9) to the nearest integer. This number represents the number of productive rehearsal hours. 10. ____________
11. Each *productive rehearsal hour* (No. 10) is worth two quality points. How many quality points do you receive for today's rehearsal? 11. ____________
12. Record the payment in your Check Register.

SEE ACTIVITY 21 FOR PRACTICE VERSION

Manager________________________ Date________________________

MILESTONE 3
FORM H

REHEARSAL

♪ : You may rent a studio for one to ten hours.

1. How many problems did you answer correctly on the quiz that was returned to you today? 1. ________
2. How many questions were asked on the quiz? 2. ________
3. Represent the part of the quiz you answered correctly as a fraction. 3. ________
4. Change the fraction in No. 3 to an equivalent decimal. 4. ________
5. Round the decimal in No. 4 to the nearest hundredth. 5. ________
6. Change the decimal in No. 5 to an equivalent percent. This number represents the percentage of productive rehearsal time. 6. ________
7. How many hours will your band rehearse today? 7. ________
8. If rehearsal studio time costs $28.25 per hour, then what will today's rehearsal cost? 8. [ ]
9. Find the number of productive rehearsal hours by multiplying total hours rehearsed (No. 7) by the percentage of productive rehearsal time (No. 6). 9. ________
10. Round the number of productive rehearsal hours (No. 9) to the nearest integer. This number represents the number of productive rehearsal hours. 10. ________
11. Each *productive rehearsal hour* (No. 10) is worth two quality points. How many quality points do you receive for today's rehearsal? 11. ________
12. Record the payment in your Check Register.

SEE ACTIVITY 21 FOR PRACTICE VERSION

Manager______________________________ Date______________________________

MILESTONE 3
FORM H

REHEARSAL

♪ : You may rent a studio for one to ten hours.

1. How many problems did you answer correctly on the quiz that was returned to you today? 1. ________

2. How many questions were asked on the quiz? 2. ________

3. Represent the part of the quiz you answered correctly as a fraction. 3. ________

4. Change the fraction in No. 3 to an equivalent decimal. 4. ________

5. Round the decimal in No. 4 to the nearest hundredth. 5. ________

6. Change the decimal in No. 5 to an equivalent percent. This number represents the percentage of productive rehearsal time. 6. ________

7. How many hours will your band rehearse today? 7. ________

8. If rehearsal studio time costs $28.25 per hour, then what will today's rehearsal cost? 8. ________

9. Find the number of productive rehearsal hours by multiplying total hours rehearsed (No. 7) by the percentage of productive rehearsal time (No. 6). 9. ________

10. Round the number of productive rehearsal hours (No. 9) to the nearest integer. This number represents the number of productive rehearsal hours. 10. ________

11. Each *productive rehearsal hour* (No. 10) is worth two quality points. How many quality points do you receive for today's rehearsal? 11. ________

12. Record the payment in your Check Register.

SEE ACTIVITY 21 FOR PRACTICE VERSION

Manager________________________________ Date________________________

MILESTONE 3
FORM H

REHEARSAL

♪ : You may rent a studio for one to ten hours.

1. How many problems did you answer correctly on the quiz that was returned to you today? 1. ____________

2. How many questions were asked on the quiz? 2. ____________

3. Represent the part of the quiz you answered correctly as a fraction. 3. ____________

4. Change the fraction in No. 3 to an equivalent decimal. 4. ____________

5. Round the decimal in No. 4 to the nearest hundredth. 5. ____________

6. Change the decimal in No. 5 to an equivalent percent. This number represents the percentage of productive rehearsal time. 6. ____________

7. How many hours will your band rehearse today? 7. ____________

8. If rehearsal studio time costs $28.25 per hour, then what will today's rehearsal cost? 8. ____________

9. Find the number of productive rehearsal hours by multiplying total hours rehearsed (No. 7) by the percentage of productive rehearsal time (No. 6). 9. ____________

10. Round the number of productive rehearsal hours (No. 9) to the nearest integer. This number represents the number of productive rehearsal hours. 10. ____________

11. Each *productive rehearsal hour* (No. 10) is worth two quality points. How many quality points do you receive for today's rehearsal? 11. ____________

12. Record the payment in your Check Register.

SEE ACTIVITY 21 FOR PRACTICE VERSION

Manager____________________ Date____________________

MILESTONE 3
FORM H

REHEARSAL

♪ : You may rent a studio for one to ten hours.

1. How many problems did you answer correctly on the quiz that was returned to you today? 1. __________

2. How many questions were asked on the quiz? 2. __________

3. Represent the part of the quiz you answered correctly as a fraction. 3. __________

4. Change the fraction in No. 3 to an equivalent decimal. 4. __________

5. Round the decimal in No. 4 to the nearest hundredth. 5. __________

6. Change the decimal in No. 5 to an equivalent percent. This number represents the percentage of productive rehearsal time. 6. __________

7. How many hours will your band rehearse today? 7. __________

8. If rehearsal studio time costs $28.25 per hour, then what will today's rehearsal cost? 8. [ ]

9. Find the number of productive rehearsal hours by multiplying total hours rehearsed (No. 7) by the percentage of productive rehearsal time (No. 6). 9. __________

10. Round the number of productive rehearsal hours (No. 9) to the nearest integer. This number represents the number of productive rehearsal hours. 10. __________

11. Each *productive rehearsal hour* (No. 10) is worth two quality points. How many quality points do you receive for today's rehearsal? 11. __________

12. Record the payment in your Check Register.

SEE ACTIVITY 21 FOR PRACTICE VERSION

Manager____________________ Date____________________

MILESTONE 3
FORM H

REHEARSAL

♪ : You may rent a studio for one to ten hours.

1. How many problems did you answer correctly on the quiz that was returned to you today? 1. __________
2. How many questions were asked on the quiz? 2. __________
3. Represent the part of the quiz you answered correctly as a fraction. 3. __________
4. Change the fraction in No. 3 to an equivalent decimal. 4. __________
5. Round the decimal in No. 4 to the nearest hundredth. 5. __________
6. Change the decimal in No. 5 to an equivalent percent. This number represents the percentage of productive rehearsal time. 6. __________
7. How many hours will your band rehearse today? 7. __________
8. If rehearsal studio time costs $28.25 per hour, then what will today's rehearsal cost? 8. \[\]
9. Find the number of productive rehearsal hours by multiplying total hours rehearsed (No. 7) by the percentage of productive rehearsal time (No. 6). 9. __________
10. Round the number of productive rehearsal hours (No. 9) to the nearest integer. This number represents the number of productive rehearsal hours. 10. __________
11. Each *productive rehearsal hour* (No. 10) is worth two quality points. How many quality points do you receive for today's rehearsal? 11. __________
12. Record the payment in your Check Register.

SEE ACTIVITY 21 FOR PRACTICE VERSION

Manager____________________ Date________________

MILESTONE 3
FORM H

REHEARSAL

♪ : You may rent a studio for one to ten hours.

1. How many problems did you answer correctly on the quiz that was returned to you today? 1. __________
2. How many questions were asked on the quiz? 2. __________
3. Represent the part of the quiz you answered correctly as a fraction. 3. __________
4. Change the fraction in No. 3 to an equivalent decimal. 4. __________
5. Round the decimal in No. 4 to the nearest hundredth. 5. __________
6. Change the decimal in No. 5 to an equivalent percent. This number represents the percentage of productive rehearsal time. 6. __________
7. How many hours will your band rehearse today? 7. __________
8. If rehearsal studio time costs $28.25 per hour, then what will today's rehearsal cost? 8. __________
9. Find the number of productive rehearsal hours by multiplying total hours rehearsed (No. 7) by the percentage of productive rehearsal time (No. 6). 9. __________
10. Round the number of productive rehearsal hours (No. 9) to the nearest integer. This number represents the number of productive rehearsal hours. 10. __________
11. Each *productive rehearsal hour* (No. 10) is worth two quality points. How many quality points do you receive for today's rehearsal? 11. __________
12. Record the payment in your Check Register.

SEE ACTIVITY 21 FOR PRACTICE VERSION

Manager____ Date________________________________

MILESTONE 3
FORM H

REHEARSAL

♪: You may rent a studio for one to ten hours.

1. How many problems did you answer correctly on the quiz that was returned to you today? 1. ____________

2. How many questions were asked on the quiz? 2. ____________

3. Represent the part of the quiz you answered correctly as a fraction. 3. ____________

4. Change the fraction in No. 3 to an equivalent decimal. 4. ____________

5. Round the decimal in No. 4 to the nearest hundredth. 5. ____________

6. Change the decimal in No. 5 to an equivalent percent. This number represents the percentage of productive rehearsal time. 6. ____________

7. How many hours will your band rehearse today? 7. ____________

8. If rehearsal studio time costs $28.25 per hour, then what will today's rehearsal cost? 8. ____________

9. Find the number of productive rehearsal hours by multiplying total hours rehearsed (No. 7) by the percentage of productive rehearsal time (No. 6). 9. ____________

10. Round the number of productive rehearsal hours (No. 9) to the nearest integer. This number represents the number of productive rehearsal hours. 10. ____________

11. Each *productive rehearsal hour* (No. 10) is worth two quality points. How many quality points do you receive for today's rehearsal? 11. ____________

12. Record the payment in your Check Register.

SEE ACTIVITY 21 FOR PRACTICE VERSION

Manager______________________ Date______________________

MILESTONE 3
FORM H

REHEARSAL

♪ : You may rent a studio for one to ten hours.

1. How many problems did you answer correctly on the quiz that was returned to you today? 1. ________
2. How many questions were asked on the quiz? 2. ________
3. Represent the part of the quiz you answered correctly as a fraction. 3. ________
4. Change the fraction in No. 3 to an equivalent decimal. 4. ________
5. Round the decimal in No. 4 to the nearest hundredth. 5. ________
6. Change the decimal in No. 5 to an equivalent percent. This number represents the percentage of productive rehearsal time. 6. ________
7. How many hours will your band rehearse today? 7. ________
8. If rehearsal studio time costs $28.25 per hour, then what will today's rehearsal cost? 8. ________
9. Find the number of productive rehearsal hours by multiplying total hours rehearsed (No. 7) by the percentage of productive rehearsal time (No. 6). 9. ________
10. Round the number of productive rehearsal hours (No. 9) to the nearest integer. This number represents the number of productive rehearsal hours. 10. ________
11. Each *productive rehearsal hour* (No. 10) is worth two quality points. How many quality points do you receive for today's rehearsal? 11. ________
12. Record the payment in your Check Register.

SEE ACTIVITY 21 FOR PRACTICE VERSION

Manager______________________ Date______________________

MILESTONE 3
FORM H

REHEARSAL

♪ : You may rent a studio for one to ten hours.

1. How many problems did you answer correctly on the quiz that was returned to you today? 1. ________
2. How many questions were asked on the quiz? 2. ________
3. Represent the part of the quiz you answered correctly as a fraction. 3. ________
4. Change the fraction in No. 3 to an equivalent decimal. 4. ________
5. Round the decimal in No. 4 to the nearest hundredth. 5. ________
6. Change the decimal in No. 5 to an equivalent percent. This number represents the percentage of productive rehearsal time. 6. ________
7. How many hours will your band rehearse today? 7. ________
8. If rehearsal studio time costs $28.25 per hour, then what will today's rehearsal cost? 8. ________
9. Find the number of productive rehearsal hours by multiplying total hours rehearsed (No. 7) by the percentage of productive rehearsal time (No. 6). 9. ________
10. Round the number of productive rehearsal hours (No. 9) to the nearest integer. This number represents the number of productive rehearsal hours. 10. ________
11. Each *productive rehearsal hour* (No. 10) is worth two quality points. How many quality points do you receive for today's rehearsal? 11. ________
12. Record the payment in your Check Register.

SEE ACTIVITY 21 FOR PRACTICE VERSION

Manager_________________________ Date_________________________

MILESTONE 3
FORM H

REHEARSAL

♪ : You may rent a studio for one to ten hours.

1. How many problems did you answer correctly on the quiz that was returned to you today? 1. ________
2. How many questions were asked on the quiz? 2. ________
3. Represent the part of the quiz you answered correctly as a fraction. 3. ________
4. Change the fraction in No. 3 to an equivalent decimal. 4. ________
5. Round the decimal in No. 4 to the nearest hundredth. 5. ________
6. Change the decimal in No. 5 to an equivalent percent. This number represents the percentage of productive rehearsal time. 6. ________
7. How many hours will your band rehearse today? 7. ________
8. If rehearsal studio time costs $28.25 per hour, then what will today's rehearsal cost? 8. ________
9. Find the number of productive rehearsal hours by multiplying total hours rehearsed (No. 7) by the percentage of productive rehearsal time (No. 6). 9. ________
10. Round the number of productive rehearsal hours (No. 9) to the nearest integer. This number represents the number of productive rehearsal hours. 10. ________
11. Each *productive rehearsal hour* (No. 10) is worth two quality points. How many quality points do you receive for today's rehearsal? 11. ________
12. Record the payment in your Check Register.

SEE ACTIVITY 21 FOR PRACTICE VERSION

MILESTONE 4

DANCE CLUBS

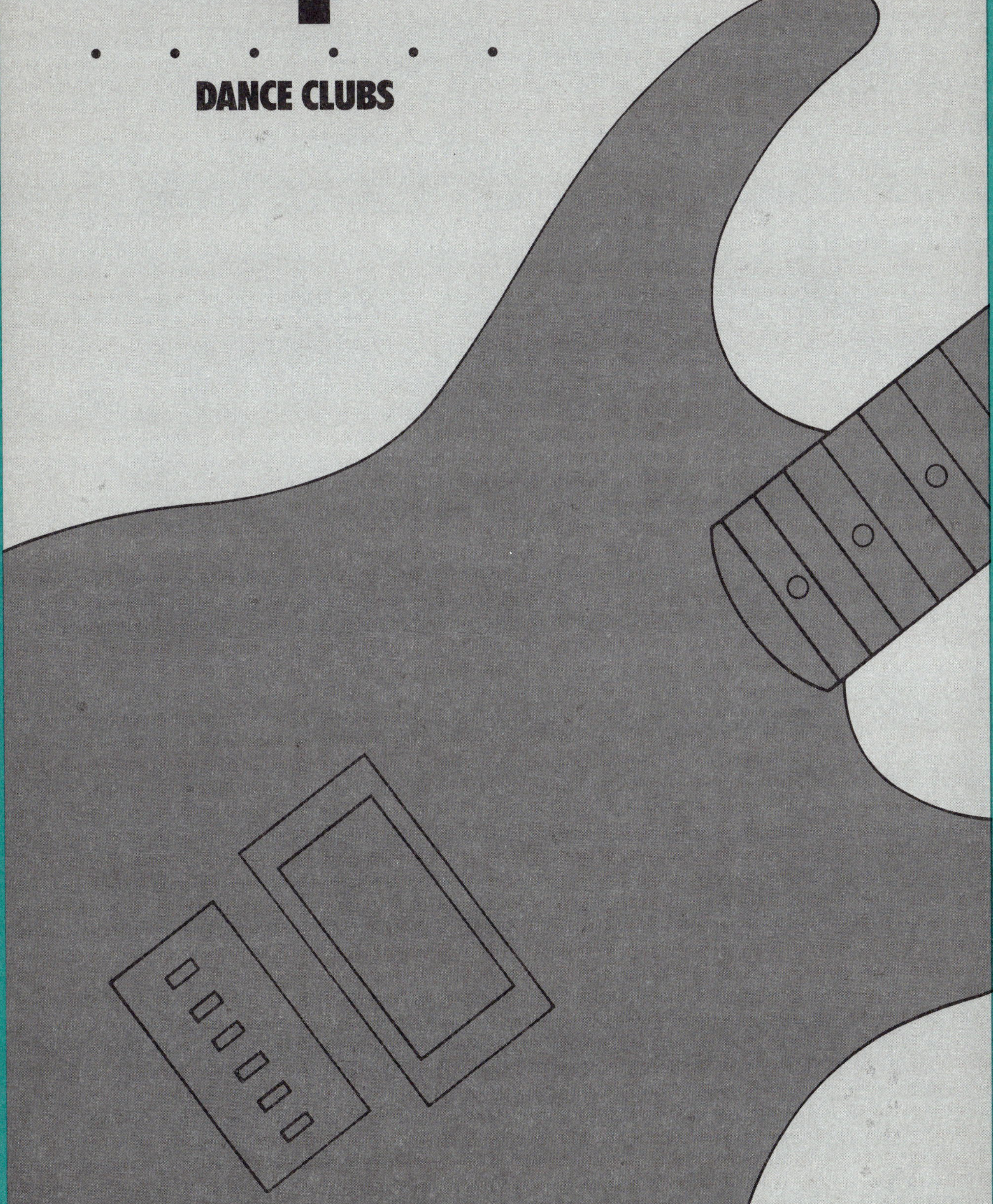

Manager________________________________ Date______________________

ACTIVITY 22

DANCE CLUB FORMS

Foundations: This is it! You're going to be earning money performing at dance clubs!

Directions: Complete the following using the information *given to you by your teacher*.

PRACTICE VERSION 1

MILESTONE 4
FORM I

DANCE CLUB

DANCE CLUB	CAPACITY (WITH TURNOVER)
THE BROOKLYN BRIDGE	1,800
CAPRICORN'S	2,200
THE CITY ZOO	3,000
GLITTER'S	900
OMNIPOP	1,200
SHARK'S	3,200
STARZ	3,100
THE THIRD RAIL	2,900

1. Where are you performing? (Use the spinner.) 1. ____________

2. What is the capacity of the dance club? (See the chart.) 2. ____________

3. How many quality points do you have? 3. ____________

4. Divide your quality points by 250. (Compute to three decimal places.) 4. ____________

5. Round your answer from No. 4 to *two* decimal places. 5. ____________

6. To determine your attendance, multiply the answer to No. 5 by the capacity (No. 2). 6. ____________

7. Each person pays $2.50 to get into the club. How much money does the club take in at the door? 7. ____________

8. You must set aside $30 for expenses such as gas, snacks, and so forth. How much money is left? 8. ____________

9. The individual shares must be computed using the following percentages:

a.	Band (60%)	9a. ______________
b.	How many band members are there?	b. ______________
c.	What is each band member's share?	c. ______________
d.	Roadies (15%)	d. ______________
e.	You have three roadies. What is each roadie's share?	e. ______________
f.	Agent (5%)	f. ______________
g.	Sound Crew (6%)	g. ______________
h.	Light Crew (4%)	h. ______________
i.	Manager (10%)	i. ______________

MILESTONE 4
FORM I

DANCE CLUB

DANCE CLUB	CAPACITY (WITH TURNOVER)
THE BROOKLYN BRIDGE	1,800
CAPRICORN'S	2,200
THE CITY ZOO	3,000
GLITTER'S	900
OMNIPOP	1,200
SHARK'S	3,200
STARZ	3,100
THE THIRD RAIL	2,900

1. Where are you performing? (Use the spinner.) 1. ____________

2. What is the capacity of the dance club? (See the chart.) 2. ____________

3. How many quality points do you have? 3. ____________

4. Divide your quality points by 250. (Compute to three decimal places.) 4. ____________

5. Round your answer from No. 4 to *two* decimal places. 5. ____________

6. To determine your attendance, multiply the answer to No. 5 by the capacity (No. 2). 6. ____________

7. Each person pays $2.50 to get into the club. How much money does the club take in at the door? 7. ____________

8. You must set aside $30 for expenses such as gas, snacks, and so forth. How much money is left? 8. ____________

9. The individual shares must be computed using the following percentages:

a.	Band (60%)	9a. __________
b.	How many band members are there?	b. __________
c.	What is each band member's share?	c. __________
d.	Roadies (15%)	d. __________
e.	You have three roadies. What is each roadie's share?	e. __________
f.	Agent (5%)	f. __________
g.	Sound Crew (6%)	g. __________
h.	Light Crew (4%)	h. __________
i.	Manager (10%)	i. __________

Manager________________________ Date________________________

MILESTONE 4
FORM I

DANCE CLUB

DANCE CLUB	CAPACITY (WITH TURNOVERS)
THE BROOKLYN BRIDGE	1,800
CAPRICORN'S	2,200
THE CITY ZOO	3,000
GLITTER'S	900
OMNIPOP	1,200
SHARK'S	3,200
STARZ	3,100
THE THIRD RAIL	2,900

1. Where are you performing? (Use the spinner.) 1. ____________

2. What is the capacity of the dance club? (See the chart.) 2. ____________

3. How many quality points do you have? 3. ____________

4. Divide your quality points by 250. (Compute to three decimal places.) 4. ____________

5. Round your answer from No. 4 to *two* decimal places. 5. ____________

6. To determine your attendance, multiply the answer to No. 5 by the capacity (No. 2). 6. ____________

7. Each person pays $2.50 to get into the club. How much money does the club take in at the door? 7. ____________

8. You must set aside $30 for expenses such as gas, snacks, and so forth. How much money is left? 8. ☐

9. The individual shares must be computed using the following percentages:

   a. Band (60%) 9a. ____________

   b. How many band members are there? b. ____________

   c. What is each band member's share? c. ____________

d. Roadies (15%) d. ________________

e. You have three roadies. What is each roadie's share? e. ________________

f. Agent (5%) f. ________________

g. Sound Crew (6%) g. ________________

h. Light Crew (4%) h. ________________

i. Manager (10%) i. ________________

10. Deposit the manager's share in your Check Register.

SEE ACTIVITY 22 FOR PRACTICE VERSION

Manager____________________ Date____________________

MILESTONE 4
FORM I

DANCE CLUB

DANCE CLUB	CAPACITY (WITH TURNOVERS)
THE BROOKLYN BRIDGE	1,800
CAPRICORN'S	2,200
THE CITY ZOO	3,000
GLITTER'S	900
OMNIPOP	1,200
SHARK'S	3,200
STARZ	3,100
THE THIRD RAIL	2,900

1. Where are you performing? (Use the spinner.) — 1. __________
2. What is the capacity of the dance club? (See the chart.) — 2. __________
3. How many quality points do you have? — 3. __________
4. Divide your quality points by 250. (Compute to three decimal places.) — 4. __________
5. Round your answer from No. 4 to *two* decimal places. — 5. __________
6. To determine your attendance, multiply the answer to No. 5 by the capacity (No. 2). — 6. __________
7. Each person pays $2.50 to get into the club. How much money does the club take in at the door? — 7. __________
8. You must set aside $30 for expenses such as gas, snacks, and so forth. How much money is left? — 8. [ ]
9. The individual shares must be computed using the following percentages:
   a. Band (60%) — 9a. __________
   b. How many band members are there? — b. __________
   c. What is each band member's share? — c. __________

d. Roadies (15%) d. ____________________

e. You have three roadies. What is each roadie's share? e. ____________________

f. Agent (5%) f. ____________________

g. Sound Crew (6%) g. ____________________

h. Light Crew (4%) h. ____________________

i. Manager (10%) i. ____________________

10. Deposit the manager's share in your Check Register.

SEE ACTIVITY 22 FOR PRACTICE VERSION

Manager_______________________________ Date_______________________________

MILESTONE 4
FORM I

DANCE CLUB

DANCE CLUB	CAPACITY (WITH TURNOVERS)
THE BROOKLYN BRIDGE	1,800
CAPRICORN'S	2,200
THE CITY ZOO	3,000
GLITTER'S	900
OMNIPOP	1,200
SHARK'S	3,200
STARZ	3,100
THE THIRD RAIL	2,900

1. Where are you performing? (Use the spinner.) 1. ____________

2. What is the capacity of the dance club? (See the chart.) 2. ____________

3. How many quality points do you have? 3. ____________

4. Divide your quality points by 250. (Compute to three decimal places.) 4. ____________

5. Round your answer from No. 4 to *two* decimal places. 5. ____________

6. To determine your attendance, multiply the answer to No. 5 by the capacity (No. 2). 6. ____________

7. Each person pays $2.50 to get into the club. How much money does the club take in at the door? 7. ____________

8. You must set aside $30 for expenses such as gas, snacks, and so forth. How much money is left? 8. ____________

9. The individual shares must be computed using the following percentages:

   a. Band (60%) 9a. ____________

   b. How many band members are there? b. ____________

   c. What is each band member's share? c. ____________

d. Roadies (15%) d. ________________

e. You have three roadies. What is each roadie's share? e. ________________

f. Agent (5%) f. ________________

g. Sound Crew (6%) g. ________________

h. Light Crew (4%) h. ________________

i. Manager (10%) i. ________________

10. Deposit the manager's share in your Check Register.

SEE ACTIVITY 22 FOR PRACTICE VERSION

Manager____________________ Date____________________

MILESTONE 4
FORM I

DANCE CLUB

DANCE CLUB	CAPACITY (WITH TURNOVERS)
THE BROOKLYN BRIDGE	1,800
CAPRICORN'S	2,200
THE CITY ZOO	3,000
GLITTER'S	900
OMNIPOP	1,200
SHARK'S	3,200
STARZ	3,100
THE THIRD RAIL	2,900

1. Where are you performing? (Use the spinner.) 1. __________

2. What is the capacity of the dance club? (See the chart.) 2. __________

3. How many quality points do you have? 3. __________

4. Divide your quality points by 250. (Compute to three decimal places.) 4. __________

5. Round your answer from No. 4 to *two* decimal places. 5. __________

6. To determine your attendance, multiply the answer to No. 5 by the capacity (No. 2). 6. __________

7. Each person pays $2.50 to get into the club. How much money does the club take in at the door? 7. __________

8. You must set aside $30 for expenses such as gas, snacks, and so forth. How much money is left? 8. [ ]

9. The individual shares must be computed using the following percentages:

   a. Band (60%) 9a. __________

   b. How many band members are there? b. __________

   c. What is each band member's share? c. __________

d. Roadies (15%) d. ____________________

e. You have three roadies. What is each roadie's share? e. ____________________

f. Agent (5%) f. ____________________

g. Sound Crew (6%) g. ____________________

h. Light Crew (4%) h. ____________________

i. Manager (10%) i. ____________________

10. Deposit the manager's share in your Check Register.

SEE ACTIVITY 22 FOR PRACTICE VERSION

Manager________________________ Date________________________

MILESTONE 4
FORM I

DANCE CLUB

DANCE CLUB	CAPACITY (WITH TURNOVERS)
THE BROOKLYN BRIDGE	1,800
CAPRICORN'S	2,200
THE CITY ZOO	3,000
GLITTER'S	900
OMNIPOP	1,200
SHARK'S	3,200
STARZ	3,100
THE THIRD RAIL	2,900

1. Where are you performing? (Use the spinner.) 1. ____________
2. What is the capacity of the dance club? (See the chart.) 2. ____________
3. How many quality points do you have? 3. ____________
4. Divide your quality points by 250. (Compute to three decimal places.) 4. ____________
5. Round your answer from No. 4 to *two* decimal places. 5. ____________
6. To determine your attendance, multiply the answer to No. 5 by the capacity (No. 2). 6. ____________
7. Each person pays $2.50 to get into the club. How much money does the club take in at the door? 7. ____________
8. You must set aside $30 for expenses such as gas, snacks, and so forth. How much money is left? 8. [ ]
9. The individual shares must be computed using the following percentages:
   a. Band (60%) 9a. ____________
   b. How many band members are there? b. ____________
   c. What is each band member's share? c. ____________

d. Roadies (15%) d. ____________________

e. You have three roadies. What is each roadie's share? e. ____________________

f. Agent (5%) f. ____________________

g. Sound Crew (6%) g. ____________________

h. Light Crew (4%) h. ____________________

i. Manager (10%) i. ____________________

10. Deposit the manager's share in your Check Register.

SEE ACTIVITY 22 FOR PRACTICE VERSION

Manager____________________ Date____________________

MILESTONE 4
FORM I

DANCE CLUB

DANCE CLUB	CAPACITY (WITH TURNOVERS)
THE BROOKLYN BRIDGE	1,800
CAPRICORN'S	2,200
THE CITY ZOO	3,000
GLITTER'S	900
OMNIPOP	1,200
SHARK'S	3,200
STARZ	3,100
THE THIRD RAIL	2,900

1. Where are you performing? (Use the spinner.) 1. __________
2. What is the capacity of the dance club? (See the chart.) 2. __________
3. How many quality points do you have? 3. __________
4. Divide your quality points by 250. (Compute to three decimal places.) 4. __________
5. Round your answer from No. 4 to *two* decimal places. 5. __________
6. To determine your attendance, multiply the answer to No. 5 by the capacity (No. 2). 6. __________
7. Each person pays $2.50 to get into the club. How much money does the club take in at the door? 7. __________
8. You must set aside $30 for expenses such as gas, snacks, and so forth. How much money is left? 8. [ ]
9. The individual shares must be computed using the following percentages:
   a. Band (60%) 9a. __________
   b. How many band members are there? b. __________
   c. What is each band member's share? c. __________

d. Roadies (15%) d. ____________

e. You have three roadies. What is each roadie's share? e. ____________

f. Agent (5%) f. ____________

g. Sound Crew (6%) g. ____________

h. Light Crew (4%) h. ____________

i. Manager (10%) i. ____________

10. Deposit the manager's share in your Check Register.

SEE ACTIVITY 22 FOR PRACTICE VERSION

Manager____________________ Date____________________

MILESTONE 4
FORM I

DANCE CLUB

DANCE CLUB	CAPACITY (WITH TURNOVERS)
THE BROOKLYN BRIDGE	1,800
CAPRICORN'S	2,200
THE CITY ZOO	3,000
GLITTER'S	900
OMNIPOP	1,200
SHARK'S	3,200
STARZ	3,100
THE THIRD RAIL	2,900

1. Where are you performing? (Use the spinner.) 1. __________
2. What is the capacity of the dance club? (See the chart.) 2. __________
3. How many quality points do you have? 3. __________
4. Divide your quality points by 250. (Compute to three decimal places.) 4. __________
5. Round your answer from No. 4 to *two* decimal places. 5. __________
6. To determine your attendance, multiply the answer to No. 5 by the capacity (No. 2). 6. __________
7. Each person pays $2.50 to get into the club. How much money does the club take in at the door? 7. __________
8. You must set aside $30 for expenses such as gas, snacks, and so forth. How much money is left? 8. __________
9. The individual shares must be computed using the following percentages:
   a. Band (60%) 9a. __________
   b. How many band members are there? b. __________
   c. What is each band member's share? c. __________

d. Roadies (15%) d. ______________

e. You have three roadies. What is each roadie's share? e. ______________

f. Agent (5%) f. ______________

g. Sound Crew (6%) g. ______________

h. Light Crew (4%) h. ______________

i. Manager (10%) i. ______________

10. Deposit the manager's share in your Check Register.

SEE ACTIVITY 22 FOR PRACTICE VERSION

Manager______________________________ Date______________________________

MILESTONE 4
FORM I

DANCE CLUB

DANCE CLUB	CAPACITY (WITH TURNOVERS)
THE BROOKLYN BRIDGE	1,800
CAPRICORN'S	2,200
THE CITY ZOO	3,000
GLITTER'S	900
OMNIPOP	1,200
SHARK'S	3,200
STARZ	3,100
THE THIRD RAIL	2,900

1. Where are you performing? (Use the spinner.) 1. ________

2. What is the capacity of the dance club? (See the chart.) 2. ________

3. How many quality points do you have? 3. ________

4. Divide your quality points by 250. (Compute to three decimal places.) 4. ________

5. Round your answer from No. 4 to *two* decimal places. 5. ________

6. To determine your attendance, multiply the answer to No. 5 by the capacity (No. 2). 6. ________

7. Each person pays $2.50 to get into the club. How much money does the club take in at the door? 7. ________

8. You must set aside $30 for expenses such as gas, snacks, and so forth. How much money is left? 8. ☐

9. The individual shares must be computed using the following percentages:

   a. Band (60%) 9a. ________

   b. How many band members are there? b. ________

   c. What is each band member's share? c. ________

d. Roadies (15%) d. ____________________

e. You have three roadies. What is each roadie's share? e. ____________________

f. Agent (5%) f. ____________________

g. Sound Crew (6%) g. ____________________

h. Light Crew (4%) h. ____________________

i. Manager (10%) i. ____________________

10. Deposit the manager's share in your Check Register.

SEE ACTIVITY 22 FOR PRACTICE VERSION

Manager______________________ Date______________________

MILESTONE 4
FORM I

DANCE CLUB

DANCE CLUB	CAPACITY (WITH TURNOVERS)
THE BROOKLYN BRIDGE	1,800
CAPRICORN'S	2,200
THE CITY ZOO	3,000
GLITTER'S	900
OMNIPOP	1,200
SHARK'S	3,200
STARZ	3,100
THE THIRD RAIL	2,900

1. Where are you performing? (Use the spinner.) 1. ______________

2. What is the capacity of the dance club? (See the chart.) 2. ______________

3. How many quality points do you have? 3. ______________

4. Divide your quality points by 250. (Compute to three decimal places.) 4. ______________

5. Round your answer from No. 4 to *two* decimal places. 5. ______________

6. To determine your attendance, multiply the answer to No. 5 by the capacity (No. 2). 6. ______________

7. Each person pays $2.50 to get into the club. How much money does the club take in at the door? 7. ______________

8. You must set aside $30 for expenses such as gas, snacks, and so forth. How much money is left? 8. ☐

9. The individual shares must be computed using the following percentages:

   a. Band (60%) 9a. ______________

   b. How many band members are there? b. ______________

   c. What is each band member's share? c. ______________

d. Roadies (15%) d. ____________________

e. You have three roadies. What is each roadie's share? e. ____________________

f. Agent (5%) f. ____________________

g. Sound Crew (6%) g. ____________________

h. Light Crew (4%) h. ____________________

i. Manager (10%) i. ____________________

10. Deposit the manager's share in your Check Register.

SEE ACTIVITY 22 FOR PRACTICE VERSION

Manager ______________________ Date ______________________

MILESTONE 4
FORM I

DANCE CLUB

DANCE CLUB	CAPACITY (WITH TURNOVERS)
THE BROOKLYN BRIDGE	1,800
CAPRICORN'S	2,200
THE CITY ZOO	3,000
GLITTER'S	900
OMNIPOP	1,200
SHARK'S	3,200
STARZ	3,100
THE THIRD RAIL	2,900

1. Where are you performing? (Use the spinner.) 1. ______________

2. What is the capacity of the dance club? (See the chart.) 2. ______________

3. How many quality points do you have? 3. ______________

4. Divide your quality points by 250. (Compute to three decimal places.) 4. ______________

5. Round your answer from No. 4 to *two* decimal places. 5. ______________

6. To determine your attendance, multiply the answer to No. 5 by the capacity (No. 2). 6. ______________

7. Each person pays $2.50 to get into the club. How much money does the club take in at the door? 7. ______________

8. You must set aside $30 for expenses such as gas, snacks, and so forth. How much money is left? 8. ☐

9. The individual shares must be computed using the following percentages:

   a. Band (60%) 9a. ______________

   b. How many band members are there? b. ______________

   c. What is each band member's share? c. ______________

d. Roadies (15%) d. ____________

e. You have three roadies. What is each roadie's share? e. ____________

f. Agent (5%) f. ____________

g. Sound Crew (6%) g. ____________

h. Light Crew (4%) h. ____________

i. Manager (10%) i. ____________

10. Deposit the manager's share in your Check Register.

SEE ACTIVITY 22 FOR PRACTICE VERSION

Manager______________________________ Date______________________________

MILESTONE 4
FORM I

DANCE CLUB

DANCE CLUB	CAPACITY (WITH TURNOVERS)
THE BROOKLYN BRIDGE	1,800
CAPRICORN'S	2,200
THE CITY ZOO	3,000
GLITTER'S	900
OMNIPOP	1,200
SHARK'S	3,200
STARZ	3,100
THE THIRD RAIL	2,900

1. Where are you performing? (Use the spinner.) 1. ______________
2. What is the capacity of the dance club? (See the chart.) 2. ______________
3. How many quality points do you have? 3. ______________
4. Divide your quality points by 250. (Compute to three decimal places.) 4. ______________
5. Round your answer from No. 4 to *two* decimal places. 5. ______________
6. To determine your attendance, multiply the answer to No. 5 by the capacity (No. 2). 6. ______________
7. Each person pays $2.50 to get into the club. How much money does the club take in at the door? 7. ______________
8. You must set aside $30 for expenses such as gas, snacks, and so forth. How much money is left? 8. [ ]
9. The individual shares must be computed using the following percentages:
   a. Band (60%) 9a. ______________
   b. How many band members are there? b. ______________
   c. What is each band member's share? c. ______________

d. Roadies (15%) d. ______________

e. You have three roadies. What is each roadie's share? e. ______________

f. Agent (5%) f. ______________

g. Sound Crew (6%) g. ______________

h. Light Crew (4%) h. ______________

i. Manager (10%) i. ______________

10. Deposit the manager's share in your Check Register.

SEE ACTIVITY 22 FOR PRACTICE VERSION

Manager____________________ Date____________________

MILESTONE 4
FORM I

DANCE CLUB

DANCE CLUB	CAPACITY (WITH TURNOVERS)
THE BROOKLYN BRIDGE	1,800
CAPRICORN'S	2,200
THE CITY ZOO	3,000
GLITTER'S	900
OMNIPOP	1,200
SHARK'S	3,200
STARZ	3,100
THE THIRD RAIL	2,900

1. Where are you performing? (Use the spinner.) 1. __________
2. What is the capacity of the dance club? (See the chart.) 2. __________
3. How many quality points do you have? 3. __________
4. Divide your quality points by 250. (Compute to three decimal places.) 4. __________
5. Round your answer from No. 4 to *two* decimal places. 5. __________
6. To determine your attendance, multiply the answer to No. 5 by the capacity (No. 2). 6. __________
7. Each person pays $2.50 to get into the club. How much money does the club take in at the door? 7. __________
8. You must set aside $30 for expenses such as gas, snacks, and so forth. How much money is left? 8. __________
9. The individual shares must be computed using the following percentages:
   a. Band (60%) 9a. __________
   b. How many band members are there? b. __________
   c. What is each band member's share? c. __________

d. Roadies (15%) d. ____________

e. You have three roadies. What is each roadie's share? e. ____________

f. Agent (5%) f. ____________

g. Sound Crew (6%) g. ____________

h. Light Crew (4%) h. ____________

i. Manager (10%) i. ____________

10. Deposit the manager's share in your Check Register.

SEE ACTIVITY 22 FOR PRACTICE VERSION

Manager____________________ Date____________________

MILESTONE 4
FORM I

DANCE CLUB

DANCE CLUB	CAPACITY (WITH TURNOVERS)
THE BROOKLYN BRIDGE	1,800
CAPRICORN'S	2,200
THE CITY ZOO	3,000
GLITTER'S	900
OMNIPOP	1,200
SHARK'S	3,200
STARZ	3,100
THE THIRD RAIL	2,900

1. Where are you performing? (Use the spinner.) 1. __________
2. What is the capacity of the dance club? (See the chart.) 2. __________
3. How many quality points do you have? 3. __________
4. Divide your quality points by 250. (Compute to three decimal places.) 4. __________
5. Round your answer from No. 4 to *two* decimal places. 5. __________
6. To determine your attendance, multiply the answer to No. 5 by the capacity (No. 2). 6. __________
7. Each person pays $2.50 to get into the club. How much money does the club take in at the door? 7. __________
8. You must set aside $30 for expenses such as gas, snacks, and so forth. How much money is left? 8. [ ]
9. The individual shares must be computed using the following percentages:
   a. Band (60%) 9a. __________
   b. How many band members are there? b. __________
   c. What is each band member's share? c. __________

d. Roadies (15%) d. ____________

e. You have three roadies. What is each roadie's share? e. ____________

f. Agent (5%) f. ____________

g. Sound Crew (6%) g. ____________

h. Light Crew (4%) h. ____________

i. Manager (10%) i. ____________

10. Deposit the manager's share in your Check Register.

SEE ACTIVITY 22 FOR PRACTICE VERSION

Manager________________________ Date________________

MILESTONE 4
FORM I

DANCE CLUB

DANCE CLUB	CAPACITY (WITH TURNOVERS)
THE BROOKLYN BRIDGE	1,800
CAPRICORN'S	2,200
THE CITY ZOO	3,000
GLITTER'S	900
OMNIPOP	1,200
SHARK'S	3,200
STARZ	3,100
THE THIRD RAIL	2,900

1. Where are you performing? (Use the spinner.) 1. ______________

2. What is the capacity of the dance club? (See the chart.) 2. ______________

3. How many quality points do you have? 3. ______________

4. Divide your quality points by 250. (Compute to three decimal places.) 4. ______________

5. Round your answer from No. 4 to *two* decimal places. 5. ______________

6. To determine your attendance, multiply the answer to No. 5 by the capacity (No. 2). 6. ______________

7. Each person pays $2.50 to get into the club. How much money does the club take in at the door? 7. ______________

8. You must set aside $30 for expenses such as gas, snacks, and so forth. How much money is left? 8. ______________

9. The individual shares must be computed using the following percentages:

   a. Band (60%) 9a. ______________

   b. How many band members are there? b. ______________

   c. What is each band member's share? c. ______________

d. Roadies (15%) d. ____________

e. You have three roadies. What is each roadie's share? e. ____________

f. Agent (5%) f. ____________

g. Sound Crew (6%) g. ____________

h. Light Crew (4%) h. ____________

i. Manager (10%) i. ____________

10. Deposit the manager's share in your Check Register.

SEE ACTIVITY 22 FOR PRACTICE VERSION

Manager____________________ Date____________________

MILESTONE 4
FORM I

DANCE CLUB

DANCE CLUB	CAPACITY (WITH TURNOVERS)
THE BROOKLYN BRIDGE	1,800
CAPRICORN'S	2,200
THE CITY ZOO	3,000
GLITTER'S	900
OMNIPOP	1,200
SHARK'S	3,200
STARZ	3,100
THE THIRD RAIL	2,900

1. Where are you performing? (Use the spinner.) 1. __________
2. What is the capacity of the dance club? (See the chart.) 2. __________
3. How many quality points do you have? 3. __________
4. Divide your quality points by 250. (Compute to three decimal places.) 4. __________
5. Round your answer from No. 4 to *two* decimal places. 5. __________
6. To determine your attendance, multiply the answer to No. 5 by the capacity (No. 2). 6. __________
7. Each person pays $2.50 to get into the club. How much money does the club take in at the door? 7. __________
8. You must set aside $30 for expenses such as gas, snacks, and so forth. How much money is left? 8. [ ]
9. The individual shares must be computed using the following percentages:

   a. Band (60%) 9a. __________

   b. How many band members are there? b. __________

   c. What is each band member's share? c. __________

d. Roadies (15%) d. ____________________

e. You have three roadies. What is each roadie's share? e. ____________________

f. Agent (5%) f. ____________________

g. Sound Crew (6%) g. ____________________

h. Light Crew (4%) h. ____________________

i. Manager (10%) i. ____________________

10. Deposit the manager's share in your Check Register.

SEE ACTIVITY 22 FOR PRACTICE VERSION

DEMO RATE SCHEDULE

STUDIO A

	per hour
MONO AND 2 TRACK RECORDING	$ 40.00
8 TRACK RECORDING AND MIXING	50.00
16 TRACK RECORDING AND MIXING	85.00
24 TRACK RECORDING AND MIXING	100.00

STUDIO INSTRUMENTS AND AMPLIFIERS ARE AVAILABLE AT NO EXTRA CHARGE

MATERIALS

1/4″ TAPE PER 2400-FOOT REEL	$ 20.00
1″ TAPE PER 2400-FOOT REEL	60.00
2″ TAPE PER 2400-FOOT REEL	140.00
C-10 CASSETTE	3.50
C-46 CASSETTE	4.50
C-90 CASSETTE	5.50
EMPTY PLASTIC REEL	1.00
EMPTY METAL REEL	5.00

PLEASE NOTE:

A DEPOSIT OF 1 HOUR STUDIO TIME IS REQUIRED TO RESERVE TIME (REFUNDABLE UPON 24 HOUR NOTICE OF CANCELLATION)

A CASH PAYMENT OR CERTIFIED CHECK IS DUE UPON COMPLETION OF EACH DAY'S SESSION UNLESS CREDIT HAS BEEN ESTABLISHED

2 YEAR TAPE STORAGE IS AVAILABLE AT CLIENT'S RISK

NO MATERIAL WILL BE RELEASED UNTIL PAYMENT HAS BEEN MADE IN FULL

Manager____________________ Date____________________

MILESTONE 4
FORM J

DEMO

As the manager, you must pay for the entire cost of a demo. Use the Demo Rate Schedule (page 137) to answer the following questions:

1. How much will 19 hours of 24 track recording and mixing cost? 1. ________________

2. How much will a reel of 2-inch tape cost? 2. ________________

3. How much will 150 "C-46" cassettes cost? 3. ________________

4. Based on Nos. 1, 2, and 3, what is the cost of the demo? 4. ________________

5. If the sales tax rate is $8\frac{1}{4}\%$, the tax is 5. ________________

6. The total cost of the demo with tax is 6. ________________

7. Record the payment in your Check Register.

Write a check payable to the Superstar Recording Studios for the demo. Make the appropriate Check Register entry.

No.

____________________ 19____ 01-1234 / 0210

PAY TO THE ORDER OF ______________________________ $ __________

____ Dollars

For Classroom Use Only

FIRST STATE BANK
10 STATE STREET
NEW YORK, NEW YORK 14614-1100

FOR ____________________ ____________________

⑆021012349⑆ 635 000 127⑈

Manager____________________ Date________________

ACTIVITY 23

SIGNED NUMBERS

Foundations: The band manager receives the mail addressed to the band. The mail includes bills (that must be paid) and money which the band has earned. These can be represented as positive and negative numbers.

Directions: "Read" the mail, and represent each piece of mail as a positive or negative number. Write the day's total finances as a positive or negative number in the space to the right.

EXAMPLE

	INCOME + $380.40	+	BILL − $689.20			= -$308.80
1.	INCOME $1,586.00	+	INCOME $111.40			= ______
2.	BILL $2,800.00	+	BILL $830.10			= ______
3.	INCOME $1,288.50	+	BILL $1,750.50			= ______
4.	INCOME $2,749.20	+	INCOME $3,220.10			= ______
5.	BILL $234.98	+	BILL $670.30			= ______
6.	BILL $3,600.00	+	BILL $2,849.30	+	INCOME $7,020.50	= ______
7.	INCOME $412.75	+	BILL $5,900.10	+	INCOME $528.57	= ______

ACTIVITY 24

SIGNED NUMBERS

Foundations: When a band is formed, each member is aware of the band's potential for success or failure. Profits and losses occur in all the milestones.

Directions: Your band is trying to increase dance club attendance by featuring one "giveaway" gimmick each month. The band members have agreed to pay for the giveaway items. In each of the following months, determine if the gimmicks resulted in a profit or loss.

Manager__________________________ Date____________________

MONTH	GIMMICK	EXPENSES	DANCE CLUB SALARY	PROFIT OR LOSS	# OF MEMBERS	PROFIT/ LOSS PER MEMBER
January	Tee Shirt Giveway	− $560.00	$700.00	+$140.00	4	+$35.00
February	Balloon Blow-Out!	− $125.50	$500.00		5	
March	Cap Craze!	− $450.10	$200.00		5	
April	Record Riot!	− $600.00	$590.00		4	
May	Bumper (sticker) Bonus	− $230.00	$230.00		5	
June	Burger Bonanza!	− $355.50	$400.95		3	
July	Sunglasses Giveway	− $200.00	$125.00		5	
August	Beach Blanket Bash!	− $250.40	$670.00		4	
September	Back to School Nite	− $50.00	$260.00		4	
October	Jack-o-Lantern Jamboree!	− $300.00	$950.00		5	
November	Musical Muffler Madness!	− $785.00	$500.00		2	
December	Santa's Surprise!	− $290.00	$1,200.00		4	

MILESTONE

5

RECORD SALES

Manager____________________ Date________________

ACTIVITY 25

RECORD SALES FORMS

Foundations: You are about to make money from the sale of your band's first single. You must keep track of the sales with a Milestone 5/Form K – Record Sales form.

Directions: Complete the following Practice Version of Milestone 5/Form K – Record Sales form using information given to you by your teacher.

♪ : Notice that record sales income is not shared with the agent, roadies, sound crew, or light crew.

PRACTICE VERSION 1

MILESTONE 5
FORM K

RECORD SALES

1. What is the name of the song featured on your single? 1. __________
2. How many quality points do you have? 2. __________
3. Multiply your quality points by 1,000. 3. __________
4. Pick a "Reach" card; what is the number on it? 4. __________
5. Your *total record sales* for the month is the sum of Nos. 3 and 4. 5. __________
6. The record company sells the records to stores for $1.10 each. How much money does the record company receive for these sales? 6. __________
7. You must make a $5,000 payment to the record company to cover the costs of producing the single. Subtract $5,000 from record sales income (No. 6). 7. [ ]
8. After making the $5,000 payment, the individual shares must be computed using the following percentages:
   - a. Band (22%) 8a. __________
   - b. How many band members are there? b. __________
   - c. What is each band member's share? c. __________
   - d. Record Company (76%) d. __________
   - e. Manager (2%) e. __________

Manager____________________ Date________________

PRACTICE VERSION 2

MILESTONE 5
FORM K

RECORD SALES

1. What is the name of the song featured on your single? 1. ________________

2. How many quality points do you have? 2. ________________

3. Multiply your quality points by 1,000. 3. ________________

4. Pick a "Reach" card; what is the number on it? 4. ________________

5. Your *total record sales* for the month is the sum of Nos. 3 and 4. 5. ________________

6. The record company sells the records to stores for $1.10 each. How much money does the record company receive for these sales? 6. ________________

7. You must make a $5,000 payment to the record company to cover the costs of producing the single. Subtract $5,000 from record sales income (No. 6). 7. [ ]

8. After making the $5,000 payment, the individual shares must be computed using the following percentages:

   a. Band (22%) 8a. ________________

   b. How many band members are there? b. ________________

   c. What is each band member's share? c. ________________

   d. Record Company (76%) d. ________________

   e. Manager (2%) e. ________________

Manager______________________________ Date______________________________

MILESTONE 5
FORM K

RECORD SALES

1. What is the name of the song featured on your single? 1. ____________

2. How many quality points do you have? 2. ____________

3. Multiply your quality points by 1,000. 3. ____________

4. Pick a "Reach" card; what is the number on it? 4. ____________

5. Your *total record sales* for the month is the sum of Nos. 3 and 4. 5. ____________

6. The record company sells the records to stores for $1.10 each. How much money does the record company receive for these sales? 6. ____________

7. You must make a $5,000 payment to the record company to cover the costs of producing the single. Subtract $5,000 from record sales income (No. 6). 7. ____________

8. After making the $5,000 payment, the individual shares must be computed using the following percentages:

   a. Band (22%) 8a. ____________

   b. How many band members are there? b. ____________

   c. What is each band member's share? c. ____________

   d. Record Company (76%) d. ____________

   e. Manager (2%) e. ____________

9. Deposit the manager's share in your Check Register.

SEE ACTIVITY 25 FOR PRACTICE VERSION

Manager________________________ Date________________________

MILESTONE 5
FORM K

RECORD SALES

1. What is the name of the song featured on your single? 1. ____________

2. How many quality points do you have? 2. ____________

3. Multiply your quality points by 1,000. 3. ____________

4. Pick a "Reach" card; what is the number on it? 4. ____________

5. Your *total record sales* for the month is the sum of Nos. 3 and 4. 5. ____________

6. The record company sells the records to stores for $1.10 each. How much money does the record company receive for these sales? 6. ____________

7. You must make a $5,000 payment to the record company to cover the costs of producing the single. Subtract $5,000 from record sales income (No. 6). 7. ☐

8. After making the $5,000 payment, the individual shares must be computed using the following percentages:

   a. Band (22%) 8a. ____________

   b. How many band members are there? b. ____________

   c. What is each band member's share? c. ____________

   d. Record Company (76%) d. ____________

   e. Manager (2%) e. ____________

9. Deposit the manager's share in your Check Register.

SEE ACTIVITY 25 FOR PRACTICE VERSION

Manager____________________ Date____________________

MILESTONE 5
FORM K

RECORD SALES

1. What is the name of the song featured on your single? 1. __________

2. How many quality points do you have? 2. __________

3. Multiply your quality points by 1,000. 3. __________

4. Pick a "Reach" card; what is the number on it? 4. __________

5. Your *total record sales* for the month is the sum of Nos. 3 and 4. 5. __________

6. The record company sells the records to stores for $1.10 each. How much money does the record company receive for these sales? 6. __________

7. You must make a $5,000 payment to the record company to cover the costs of producing the single. Subtract $5,000 from record sales income (No. 6). 7. \[\]

8. After making the $5,000 payment, the individual shares must be computed using the following percentages:

   a. Band (22%) 8a. __________

   b. How many band members are there? b. __________

   c. What is each band member's share? c. __________

   d. Record Company (76%) d. __________

   e. Manager (2%) e. __________

9. Deposit the manager's share in your Check Register.

SEE ACTIVITY 25 FOR PRACTICE VERSION

Manager________________________ Date________________________

MILESTONE 5
FORM K

RECORD SALES

1. What is the name of the song featured on your single? 1. ________________

2. How many quality points do you have? 2. ________________

3. Multiply your quality points by 1,000. 3. ________________

4. Pick a "Reach" card; what is the number on it? 4. ________________

5. Your *total record sales* for the month is the sum of Nos. 3 and 4. 5. ________________

6. The record company sells the records to stores for $1.10 each. How much money does the record company receive for these sales? 6. ________________

7. You must make a $5,000 payment to the record company to cover the costs of producing the single. Subtract $5,000 from record sales income (No. 6). 7. [ ]

8. After making the $5,000 payment, the individual shares must be computed using the following percentages:

   a. Band (22%) 8a. ________________

   b. How many band members are there? b. ________________

   c. What is each band member's share? c. ________________

   d. Record Company (76%) d. ________________

   e. Manager (2%) e. ________________

9. Deposit the manager's share in your Check Register.

SEE ACTIVITY 25 FOR PRACTICE VERSION

Manager________________________ Date________________________

MILESTONE 5
FORM K

RECORD SALES

1. What is the name of the song featured on your single? 1. ____________

2. How many quality points do you have? 2. ____________

3. Multiply your quality points by 1,000. 3. ____________

4. Pick a "Reach" card; what is the number on it? 4. ____________

5. Your *total record sales* for the month is the sum of Nos. 3 and 4. 5. ____________

6. The record company sells the records to stores for $1.10 each. How much money does the record company receive for these sales? 6. ____________

7. You must make a $5,000 payment to the record company to cover the costs of producing the single. Subtract $5,000 from record sales income (No. 6). 7. [ ]

8. After making the $5,000 payment, the individual shares must be computed using the following percentages:

   a. Band (22%) 8a. ____________

   b. How many band members are there? b. ____________

   c. What is each band member's share? c. ____________

   d. Record Company (76%) d. ____________

   e. Manager (2%) e. ____________

9. Deposit the manager's share in your Check Register.

SEE ACTIVITY 25 FOR PRACTICE VERSION

Manager______________________________ Date______________________________

MILESTONE 5
FORM K

RECORD SALES

1. What is the name of the song featured on your single? 1. ____________

2. How many quality points do you have? 2. ____________

3. Multiply your quality points by 1,000. 3. ____________

4. Pick a "Reach" card; what is the number on it? 4. ____________

5. Your *total record sales* for the month is the sum of Nos. 3 and 4. 5. ____________

6. The record company sells the records to stores for $1.10 each. How much money does the record company receive for these sales? 6. ____________

7. You must make a $5,000 payment to the record company to cover the costs of producing the single. Subtract $5,000 from record sales income (No. 6). 7. ____________

8. After making the $5,000 payment, the individual shares must be computed using the following percentages:

   a. Band (22%) 8a. ____________

   b. How many band members are there? b. ____________

   c. What is each band member's share? c. ____________

   d. Record Company (76%) d. ____________

   e. Manager (2%) e. ____________

9. Deposit the manager's share in your Check Register.

SEE ACTIVITY 25 FOR PRACTICE VERSION

Manager________________________ Date________________________

MILESTONE 5
FORM K

RECORD SALES

1. What is the name of the song featured on your single? 1. ____________

2. How many quality points do you have? 2. ____________

3. Multiply your quality points by 1,000. 3. ____________

4. Pick a "Reach" card; what is the number on it? 4. ____________

5. Your *total record sales* for the month is the sum of Nos. 3 and 4. 5. ____________

6. The record company sells the records to stores for $1.10 each. How much money does the record company receive for these sales? 6. ____________

7. You must make a $5,000 payment to the record company to cover the costs of producing the single. Subtract $5,000 from record sales income (No. 6). 7. [ ]

8. After making the $5,000 payment, the individual shares must be computed using the following percentages:

   a. Band (22%) 8a. ____________

   b. How many band members are there? b. ____________

   c. What is each band member's share? c. ____________

   d. Record Company (76%) d. ____________

   e. Manager (2%) e. ____________

9. Deposit the manager's share in your Check Register.

SEE ACTIVITY 25 FOR PRACTICE VERSION

Manager________________________________ Date________________________

MILESTONE 5
FORM K

RECORD SALES

1. What is the name of the song featured on your single? 1. ______________
2. How many quality points do you have? 2. ______________
3. Multiply your quality points by 1,000. 3. ______________
4. Pick a "Reach" card; what is the number on it? 4. ______________
5. Your *total record sales* for the month is the sum of Nos. 3 and 4. 5. ______________
6. The record company sells the records to stores for $1.10 each. How much money does the record company receive for these sales? 6. ______________
7. You must make a $5,000 payment to the record company to cover the costs of producing the single. Subtract $5,000 from record sales income (No. 6). 7. [ ]
8. After making the $5,000 payment, the individual shares must be computed using the following percentages:
   a. Band (22%) 8a. ______________
   b. How many band members are there? b. ______________
   c. What is each band member's share? c. ______________
   d. Record Company (76%) d. ______________
   e. Manager (2%) e. ______________
9. Deposit the manager's share in your Check Register.

SEE ACTIVITY 25 FOR PRACTICE VERSION

Manager______________________________ Date______________________________

MILESTONE 5
FORM K

RECORD SALES

1. What is the name of the song featured on your single? 1. ____________
2. How many quality points do you have? 2. ____________
3. Multiply your quality points by 1,000. 3. ____________
4. Pick a "Reach" card; what is the number on it? 4. ____________
5. Your *total record sales* for the month is the sum of Nos. 3 and 4. 5. ____________
6. The record company sells the records to stores for $1.10 each. How much money does the record company receive for these sales? 6. ____________
7. You must make a $5,000 payment to the record company to cover the costs of producing the single. Subtract $5,000 from record sales income (No. 6). 7. ____________
8. After making the $5,000 payment, the individual shares must be computed using the following percentages:
   - a. Band (22%) 8a. ____________
   - b. How many band members are there? b. ____________
   - c. What is each band member's share? c. ____________
   - d. Record Company (76%) d. ____________
   - e. Manager (2%) e. ____________
9. Deposit the manager's share in your Check Register.

SEE ACTIVITY 25 FOR PRACTICE VERSION

Manager________________________ Date________________________

MILESTONE 5
FORM K

RECORD SALES

1. What is the name of the song featured on your single? 1. ________
2. How many quality points do you have? 2. ________
3. Multiply your quality points by 1,000. 3. ________
4. Pick a "Reach" card; what is the number on it? 4. ________
5. Your *total record sales* for the month is the sum of Nos. 3 and 4. 5. ________
6. The record company sells the records to stores for $1.10 each. How much money does the record company receive for these sales? 6. ________
7. You must make a $5,000 payment to the record company to cover the costs of producing the single. Subtract $5,000 from record sales income (No. 6). 7. ________
8. After making the $5,000 payment, the individual shares must be computed using the following percentages:
   a. Band (22%) 8a. ________
   b. How many band members are there? b. ________
   c. What is each band member's share? c. ________
   d. Record Company (76%) d. ________
   e. Manager (2%) e. ________
9. Deposit the manager's share in your Check Register.

SEE ACTIVITY 25 FOR PRACTICE VERSION

Manager________________________ Date________________________

MILESTONE 5
FORM K

RECORD SALES

1. What is the name of the song featured on your single? 1. ________

2. How many quality points do you have? 2. ________

3. Multiply your quality points by 1,000. 3. ________

4. Pick a "Reach" card; what is the number on it? 4. ________

5. Your *total record sales* for the month is the sum of Nos. 3 and 4. 5. ________

6. The record company sells the records to stores for $1.10 each. How much money does the record company receive for these sales? 6. ________

7. You must make a $5,000 payment to the record company to cover the costs of producing the single. Subtract $5,000 from record sales income (No. 6). 7. ☐

8. After making the $5,000 payment, the individual shares must be computed using the following percentages:

   a. Band (22%) 8a. ________

   b. How many band members are there? b. ________

   c. What is each band member's share? c. ________

   d. Record Company (76%) d. ________

   e. Manager (2%) e. ________

9. Deposit the manager's share in your Check Register.

SEE ACTIVITY 25 FOR PRACTICE VERSION

Manager________________________________ Date________________________________

MILESTONE 5
FORM K

RECORD SALES

1. What is the name of the song featured on your single? 1. ____________
2. How many quality points do you have? 2. ____________
3. Multiply your quality points by 1,000. 3. ____________
4. Pick a "Reach" card; what is the number on it? 4. ____________
5. Your *total record sales* for the month is the sum of Nos. 3 and 4. 5. ____________
6. The record company sells the records to stores for $1.10 each. How much money does the record company receive for these sales? 6. ____________
7. You must make a $5,000 payment to the record company to cover the costs of producing the single. Subtract $5,000 from record sales income (No. 6). 7. [ ]
8. After making the $5,000 payment, the individual shares must be computed using the following percentages:
   a. Band (22%) 8a. ____________
   b. How many band members are there? b. ____________
   c. What is each band member's share? c. ____________
   d. Record Company (76%) d. ____________
   e. Manager (2%) e. ____________
9. Deposit the manager's share in your Check Register.

SEE ACTIVITY 25 FOR PRACTICE VERSION

Manager______________________________ Date______________________________

MILESTONE 5
FORM K

RECORD SALES

1. What is the name of the song featured on your single? 1. __________

2. How many quality points do you have? 2. __________

3. Multiply your quality points by 1,000. 3. __________

4. Pick a "Reach" card; what is the number on it? 4. __________

5. Your *total record sales* for the month is the sum of Nos. 3 and 4. 5. __________

6. The record company sells the records to stores for $1.10 each. How much money does the record company receive for these sales? 6. __________

7. You must make a $5,000 payment to the record company to cover the costs of producing the single. Subtract $5,000 from record sales income (No. 6). 7. [ ]

8. After making the $5,000 payment, the individual shares must be computed using the following percentages:

   a. Band (22%) 8a. __________

   b. How many band members are there? b. __________

   c. What is each band member's share? c. __________

   d. Record Company (76%) d. __________

   e. Manager (2%) e. __________

9. Deposit the manager's share in your Check Register.

SEE ACTIVITY 25 FOR PRACTICE VERSION

Manager________________________ Date________________________

MILESTONE 5
FORM K

RECORD SALES

1. What is the name of the song featured on your single? 1. ________

2. How many quality points do you have? 2. ________

3. Multiply your quality points by 1,000. 3. ________

4. Pick a "Reach" card; what is the number on it? 4. ________

5. Your *total record sales* for the month is the sum of Nos. 3 and 4. 5. ________

6. The record company sells the records to stores for $1.10 each. How much money does the record company receive for these sales? 6. ________

7. You must make a $5,000 payment to the record company to cover the costs of producing the single. Subtract $5,000 from record sales income (No. 6). 7. ☐

8. After making the $5,000 payment, the individual shares must be computed using the following percentages:

   a. Band (22%) 8a. ________

   b. How many band members are there? b. ________

   c. What is each band member's share? c. ________

   d. Record Company (76%) d. ________

   e. Manager (2%) e. ________

9. Deposit the manager's share in your Check Register.

SEE ACTIVITY 25 FOR PRACTICE VERSION

Manager____________________ Date____________________

MILESTONE 5
FORM K

RECORD SALES

1. What is the name of the song featured on your single? 1. __________

2. How many quality points do you have? 2. __________

3. Multiply your quality points by 1,000. 3. __________

4. Pick a "Reach" card; what is the number on it? 4. __________

5. Your *total record sales* for the month is the sum of Nos. 3 and 4. 5. __________

6. The record company sells the records to stores for $1.10 each. How much money does the record company receive for these sales? 6. __________

7. You must make a $5,000 payment to the record company to cover the costs of producing the single. Subtract $5,000 from record sales income (No. 6). 7. [ ]

8. After making the $5,000 payment, the individual shares must be computed using the following percentages:

   a. Band (22%) 8a. __________

   b. How many band members are there? b. __________

   c. What is each band member's share? c. __________

   d. Record Company (76%) d. __________

   e. Manager (2%) e. __________

9. Deposit the manager's share in your Check Register.

SEE ACTIVITY 25 FOR PRACTICE VERSION

Manager____________________ Date________________

ACTIVITY 26

CONSTRUCTING BAR GRAPHS

Foundations: During business meetings, bar graphs are often used as a visual way to display information. As band manager, you will display information to record companies, radio stations, and music magazines.

Directions: On two sheets of graph paper, use the following information to construct the bar graphs for Numbers 1 and 2.

♪ : Remember to label the axes.

1. Rehearsal hours over a one-week period:

DAY	HOURS
Monday	$6\frac{1}{2}$
Tuesday	4
Wednesday	10
Thursday	$7\frac{1}{2}$
Friday	$9\frac{3}{4}$
Saturday	10
Sunday	2

2. Dance clubs and their capacities:

CLUB	CAPACITY
The Brooklyn Bridge	1,800
Capricorn's	2,200
The City Zoo	3,000
Glitter's	900
Omnipop	1,200
Shark's	3,200
Starz	3,100
The Third Rail	2,900

Manager____________________ Date____________________

ACTIVITY 27

CONSTRUCTING LINE GRAPHS

Foundations: Line graphs are used when you want to display information that changes over a period of time. As band manager, you are constantly watching your band's progress over time.

Directions: On two sheets of graph paper, use the following information to construct the line graphs for Numbers 1 and 2.

♪ : Remember to label the axes.

1. The following is a listing of payments a band manager made over the last ten days.

DAY	PAYMENT
1	$5,500
2	$ 700
3	$ 450
4	$1,700
5	$3,000
6	$4,400
7	$2,100
8	$ 150
9	$ 150
10	$ 150

2. The following is a list of the number of records sold nationally during a week.

Monday	100,000
Tuesday	120,000
Wednesday	80,000
Thursday	15,000
Friday	10,000
Saturday	5,000

Manager____________________________ Date______________________

ACTIVITY 28

CONSTRUCTING PICTOGRAPHS

Foundations: Pictographs may be interpreted quickly and are often used in magazines in the music industry. As a band manager, you use pictographs to show amounts of information quickly.

Directions: On two sheets of unlined paper, use the following information to construct the pictographs for Numbers 1 and 2.

1. Your band just made a video. Each TV set represents three airings of your video on a famous video channel.

DAY	AIRINGS
Monday	15
Tuesday	12
Wednesday	3
Thursday	7
Friday	13
Saturday	6
Sunday	10

2. Your accountant is reviewing your record sales for last year. Each record represents 10,000 records sold.

January–February	20,000
March–April	85,000
May–June	100,000
July–August	25,000
September–October	95,000
November–December	40,000

MILESTONE 6

AIRPLAY

ACTIVITY 29

AIRPLAY FORM

Foundations: Your recently released single will now be played nationally on the radio. Each time it is played, you receive $.03; this money is called a *royalty*.

Directions: Using the information given to you by your teacher, complete the Practice Version of Milestone 6/Form L – Airplay.

Manager____________________ Date________________

PRACTICE VERSION

MILESTONE 6
FORM L

AIRPLAY

1. What is the name of your band's single that is featured on the radio? 1. ________________
2. How many quality points do you have? 2. ________________
3. Square your quality points. 3. ________________
4. Pick a "Reach" card; what is the number on it? 4. ________________
5. Add your answer from No. 4 to the squared quality points (No. 3). *These are your airplays for the month.* 5. ________________
6. *Royalties:* You receive three cents ($.03) for each airplay. Compute your royalties. 6. [ ]
7. The individual royalties must be computed using the following percentages:

   a. Band (70%) 7a. ________________

   b. How many band members are there? b. ________________

   c. What is each band member's share? c. ________________

   d. Agent (12%) d. ________________

   e. Manager (18%) e. ________________

Manager____________________ Date____________________

MILESTONE 6
FORM L

AIRPLAY

1. What is the name of your band's single that is featured on the radio? 1. __________

2. How many quality points do you have? 2. __________

3. Square your quality points. 3. __________

4. Pick a "Reach" card; what is the number on it? 4. __________

5. Add your answer from No. 4 to the squared quality points (No. 3). *These are your airplays for the month.* 5. __________

6. *Royalties:* You receive three cents ($.03) for each airplay. Compute your royalties. 6. [ ]

7. The individual royalties must be computed using the following percentages:

   a. Band (70%) 7a. __________

   b. How many band members are there? b. __________

   c. What is each band member's share? c. __________

   d. Agent (12%) d. __________

   e. Manager (18%) e. __________

8. Deposit the manager's share in your Check Register.

SEE ACTIVITY 29 FOR PRACTICE VERSION

Manager____________________ Date____________________

MILESTONE 6
FORM L

AIRPLAY

1. What is the name of your band's single that is featured on the radio? 1. __________
2. How many quality points do you have? 2. __________
3. Square your quality points. 3. __________
4. Pick a "Reach" card; what is the number on it? 4. __________
5. Add your answer from No. 4 to the squared quality points (No. 3). *These are your airplays for the month.* 5. __________
6. *Royalties:* You receive three cents ($.03) for each airplay. Compute your royalties. 6. [ ]
7. The individual royalties must be computed using the following percentages:
   a. Band (70%) 7a. __________
   b. How many band members are there? b. __________
   c. What is each band member's share? c. __________
   d. Agent (12%) d. __________
   e. Manager (18%) e. __________
8. Deposit the manager's share in your Check Register.

SEE ACTIVITY 29 FOR PRACTICE VERSION

Manager________________ Date________________

MILESTONE 6
FORM L

AIRPLAY

1. What is the name of your band's single that is featured on the radio? 1. __________

2. How many quality points do you have? 2. __________

3. Square your quality points. 3. __________

4. Pick a "Reach" card; what is the number on it? 4. __________

5. Add your answer from No. 4 to the squared quality points (No. 3). *These are your airplays for the month.* 5. __________

6. *Royalties:* You receive three cents ($.03) for each airplay. Compute your royalties. 6. __________

7. The individual royalties must be computed using the following percentages:

   a. Band (70%) 7a. __________

   b. How many band members are there? b. __________

   c. What is each band member's share? c. __________

   d. Agent (12%) d. __________

   e. Manager (18%) e. __________

8. Deposit the manager's share in your Check Register.

SEE ACTIVITY 29 FOR PRACTICE VERSION

Manager__________________ Date__________________

MILESTONE 6
FORM L

AIRPLAY

1. What is the name of your band's single that is featured on the radio? 1. __________

2. How many quality points do you have? 2. __________

3. Square your quality points. 3. __________

4. Pick a "Reach" card; what is the number on it? 4. __________

5. Add your answer from No. 4 to the squared quality points (No. 3). *These are your airplays for the month.* 5. __________

6. *Royalties:* You receive three cents ($.03) for each airplay. Compute your royalties. 6. [ ]

7. The individual royalties must be computed using the following percentages:

   a. Band (70%) 7a. __________

   b. How many band members are there? b. __________

   c. What is each band member's share? c. __________

   d. Agent (12%) d. __________

   e. Manager (18%) e. __________

8. Deposit the manager's share in your Check Register.

SEE ACTIVITY 29 FOR PRACTICE VERSION

Manager________________________________ Date________________________________

MILESTONE 6
FORM L

AIRPLAY

1. What is the name of your band's single that is featured on the radio? 1. ______________
2. How many quality points do you have? 2. ______________
3. Square your quality points. 3. ______________
4. Pick a "Reach" card; what is the number on it? 4. ______________
5. Add your answer from No. 4 to the squared quality points (No. 3). *These are your airplays for the month.* 5. ______________
6. *Royalties:* You receive three cents ($.03) for each airplay. Compute your royalties. 6. ☐
7. The individual royalties must be computed using the following percentages:
   a. Band (70%) 7a. ______________
   b. How many band members are there? b. ______________
   c. What is each band member's share? c. ______________
   d. Agent (12%) d. ______________
   e. Manager (18%) e. ______________
8. Deposit the manager's share in your Check Register.

SEE ACTIVITY 29 FOR PRACTICE VERSION

Manager______________________________ Date______________________________

MILESTONE 6
FORM L

AIRPLAY

1. What is the name of your band's single that is featured on the radio? 1. ____________

2. How many quality points do you have? 2. ____________

3. Square your quality points. 3. ____________

4. Pick a "Reach" card; what is the number on it? 4. ____________

5. Add your answer from No. 4 to the squared quality points (No. 3). *These are your airplays for the month.* 5. ____________

6. *Royalties:* You receive three cents ($.03) for each airplay. Compute your royalties. 6. [ ]

7. The individual royalties must be computed using the following percentages:

   a. Band (70%) 7a. ____________

   b. How many band members are there? b. ____________

   c. What is each band member's share? c. ____________

   d. Agent (12%) d. ____________

   e. Manager (18%) e. ____________

8. Deposit the manager's share in your Check Register.

SEE ACTIVITY 29 FOR PRACTICE VERSION

Manager________________________ Date________________________

MILESTONE 6
FORM L

AIRPLAY

1. What is the name of your band's single that is featured on the radio? 1. ______________

2. How many quality points do you have? 2. ______________

3. Square your quality points. 3. ______________

4. Pick a "Reach" card; what is the number on it? 4. ______________

5. Add your answer from No. 4 to the squared quality points (No. 3). *These are your airplays for the month.* 5. ______________

6. *Royalties:* You receive three cents ($.03) for each airplay. Compute your royalties. 6. ______________

7. The individual royalties must be computed using the following percentages:

   a. Band (70%) 7a. ______________

   b. How many band members are there? b. ______________

   c. What is each band member's share? c. ______________

   d. Agent (12%) d. ______________

   e. Manager (18%) e. ______________

8. Deposit the manager's share in your Check Register.

SEE ACTIVITY 29 FOR PRACTICE VERSION

Manager____________________ Date____________________

MILESTONE 6
FORM L

AIRPLAY

1. What is the name of your band's single that is featured on the radio? 1. __________

2. How many quality points do you have? 2. __________

3. Square your quality points. 3. __________

4. Pick a "Reach" card; what is the number on it? 4. __________

5. Add your answer from No. 4 to the squared quality points (No. 3). *These are your airplays for the month.* 5. __________

6. *Royalties:* You receive three cents ($.03) for each airplay. Compute your royalties. 6. [ ]

7. The individual royalties must be computed using the following percentages:

   a. Band (70%) 7a. __________

   b. How many band members are there? b. __________

   c. What is each band member's share? c. __________

   d. Agent (12%) d. __________

   e. Manager (18%) e. __________

8. Deposit the manager's share in your Check Register.

SEE ACTIVITY 29 FOR PRACTICE VERSION

Manager________________________________ Date________________________________

MILESTONE 6
FORM L

AIRPLAY

1. What is the name of your band's single that is featured on the radio? 1. ____________

2. How many quality points do you have? 2. ____________

3. Square your quality points. 3. ____________

4. Pick a "Reach" card; what is the number on it? 4. ____________

5. Add your answer from No. 4 to the squared quality points (No. 3). *These are your airplays for the month.* 5. ____________

6. *Royalties:* You receive three cents ($.03) for each airplay. Compute your royalties. 6. [ ]

7. The individual royalties must be computed using the following percentages:

   a. Band (70%) 7a. ____________

   b. How many band members are there? b. ____________

   c. What is each band member's share? c. ____________

   d. Agent (12%) d. ____________

   e. Manager (18%) e. ____________

8. Deposit the manager's share in your Check Register.

SEE ACTIVITY 29 FOR PRACTICE VERSION

Manager________________________________ Date________________________________

MILESTONE 6
FORM L

AIRPLAY

1. What is the name of your band's single that is featured on the radio? 1. ____________

2. How many quality points do you have? 2. ____________

3. Square your quality points. 3. ____________

4. Pick a "Reach" card; what is the number on it? 4. ____________

5. Add your answer from No. 4 to the squared quality points (No. 3). *These are your airplays for the month.* 5. ____________

6. *Royalties:* You receive three cents ($.03) for each airplay. Compute your royalties. 6. [ ]

7. The individual royalties must be computed using the following percentages:

   a. Band (70%) 7a. ____________

   b. How many band members are there? b. ____________

   c. What is each band member's share? c. ____________

   d. Agent (12%) d. ____________

   e. Manager (18%) e. ____________

8. Deposit the manager's share in your Check Register.

SEE ACTIVITY 29 FOR PRACTICE VERSION

Manager____ Date________________________________

MILESTONE 6
FORM L

AIRPLAY

1. What is the name of your band's single that is featured on the radio? 1. ______________

2. How many quality points do you have? 2. ______________

3. Square your quality points. 3. ______________

4. Pick a "Reach" card; what is the number on it? 4. ______________

5. Add your answer from No. 4 to the squared quality points (No. 3). *These are your airplays for the month.* 5. ______________

6. *Royalties:* You receive three cents ($.03) for each airplay. Compute your royalties. 6. ______________

7. The individual royalties must be computed using the following percentages:

   a. Band (70%) 7a. ______________

   b. How many band members are there? b. ______________

   c. What is each band member's share? c. ______________

   d. Agent (12%) d. ______________

   e. Manager (18%) e. ______________

8. Deposit the manager's share in your Check Register.

SEE ACTIVITY 29 FOR PRACTICE VERSION

Manager________________________ Date________________________

MILESTONE 6
FORM L

AIRPLAY

1. What is the name of your band's single that is featured on the radio? 1. __________

2. How many quality points do you have? 2. __________

3. Square your quality points. 3. __________

4. Pick a "Reach" card; what is the number on it? 4. __________

5. Add your answer from No. 4 to the squared quality points (No. 3). *These are your airplays for the month.* 5. __________

6. *Royalties:* You receive three cents ($.03) for each airplay. Compute your royalties. 6. [ ]

7. The individual royalties must be computed using the following percentages:

   a. Band (70%) 7a. __________

   b. How many band members are there? b. __________

   c. What is each band member's share? c. __________

   d. Agent (12%) d. __________

   e. Manager (18%) e. __________

8. Deposit the manager's share in your Check Register.

SEE ACTIVITY 29 FOR PRACTICE VERSION

Manager____________________ Date____________________

MILESTONE 6
FORM L

AIRPLAY

1. What is the name of your band's single that is featured on the radio? 1. __________

2. How many quality points do you have? 2. __________

3. Square your quality points. 3. __________

4. Pick a "Reach" card; what is the number on it? 4. __________

5. Add your answer from No. 4 to the squared quality points (No. 3). *These are your airplays for the month.* 5. __________

6. *Royalties:* You receive three cents ($.03) for each airplay. Compute your royalties. 6. [ ]

7. The individual royalties must be computed using the following percentages:

   a. Band (70%) 7a. __________

   b. How many band members are there? b. __________

   c. What is each band member's share? c. __________

   d. Agent (12%) d. __________

   e. Manager (18%) e. __________

8. Deposit the manager's share in your Check Register.

SEE ACTIVITY 29 FOR PRACTICE VERSION

Manager________________________ Date________________________

MILESTONE 6
FORM L

AIRPLAY

1. What is the name of your band's single that is featured on the radio? 1. ________
2. How many quality points do you have? 2. ________
3. Square your quality points. 3. ________
4. Pick a "Reach" card; what is the number on it? 4. ________
5. Add your answer from No. 4 to the squared quality points (No. 3). *These are your airplays for the month.* 5. ________
6. *Royalties:* You receive three cents ($.03) for each airplay. Compute your royalties. 6. ________
7. The individual royalties must be computed using the following percentages:
   a. Band (70%) 7a. ________
   b. How many band members are there? b. ________
   c. What is each band member's share? c. ________
   d. Agent (12%) d. ________
   e. Manager (18%) e. ________
8. Deposit the manager's share in your Check Register.

SEE ACTIVITY 29 FOR PRACTICE VERSION

Manager____________________ Date____________________

MILESTONE 6
FORM L

AIRPLAY

1. What is the name of your band's single that is featured on the radio? 1. ____________

2. How many quality points do you have? 2. ____________

3. Square your quality points. 3. ____________

4. Pick a "Reach" card; what is the number on it? 4. ____________

5. Add your answer from No. 4 to the squared quality points (No. 3). *These are your airplays for the month.* 5. ____________

6. *Royalties:* You receive three cents ($.03) for each airplay. Compute your royalties. 6. [ ]

7. The individual royalties must be computed using the following percentages:

   a. Band (70%) 7a. ____________

   b. How many band members are there? b. ____________

   c. What is each band member's share? c. ____________

   d. Agent (12%) d. ____________

   e. Manager (18%) e. ____________

8. Deposit the manager's share in your Check Register.

SEE ACTIVITY 29 FOR PRACTICE VERSION

MILESTONE 7

PUBLICITY

Manager____________________________ Date____________________

ACTIVITY 30
FRACTIONS

Foundations: As band managers, you must be concerned with the stage setup before a big concert. You must have accurate measurement and fraction skills to plan equipment correctly.

Directions: Each of the following equipment setup problems needs to be solved. Can you help?

♪ : Read the problem carefully to decide whether the problem involves addition, subtraction, multiplication, or division.

1. At concerts, speaker cabinets are usually "stacked" (one placed on top of another) to save space. Here are the heights of three speaker cabinets.

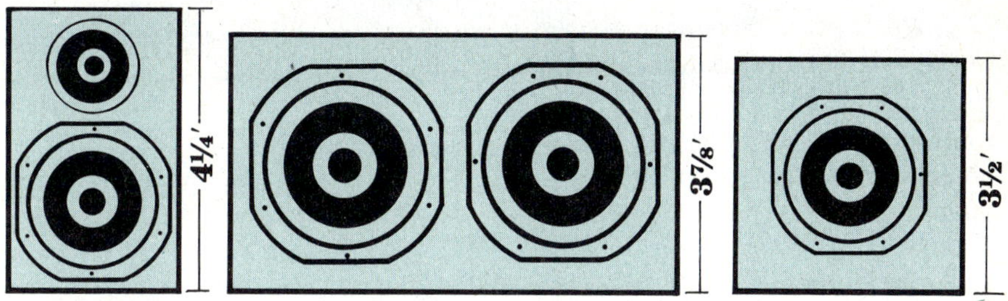

"BOSS" **"DUO-SOUND"** **"LONE STAR"**

a. What is the total height of the stack when one "Boss" is placed on top of one "Duo-Sound"? 1a. ________________

b. What is the total height when all three cabinets are stacked? b. ________________

c. How much taller is one "Boss" than one "Lone Star"? c. ________________

d. The ceiling in Glitter's is 11 feet high. Could you stack one "Lone Star" cabinet on top of two stacked "Duo-Sounds"? d. ________________

e. Which stack is taller, four "Lone Stars" or two "Duo-Sounds"? e. ________________

2. The roadies purchased a spool of microphone wire. The spool contains $67\frac{1}{2}$ feet of wire. The roadies are going to cut the wire into several $7\frac{1}{2}$-foot lengths. How many $7\frac{1}{2}$-foot lengths can they cut from the spool of $67\frac{1}{2}$ feet?

2. ________________

3. There were 640 people at Omnipop's "College Night." You found out that $\frac{7}{8}$ of the people were from Queens College. How many people from Queens College were at Omnipop's "College Night"?

3. ________________

4. The owners of Shark's are planning to buy a new curtain for the rear of the stage. Shark's has a 22-foot-high ceiling. If the stage is $5\frac{3}{4}$ feet off the ground, what is the height of the curtain?

4. ________________

Manager________________ Date________________

ACTIVITY 31
THE RULER

Foundations: Careful and exact measurements are important in the construction of musical instruments. You could not design or fix an instrument without a ruler.

Directions: To the right is a diagram of a guitar: Look at the frets. Frets are straight pieces of metal glued to the guitar neck. Different notes can be created by pressing the strings between different frets. Frets must be accurately spaced. Use your ruler skills to find the distance between the frets on the following guitars.

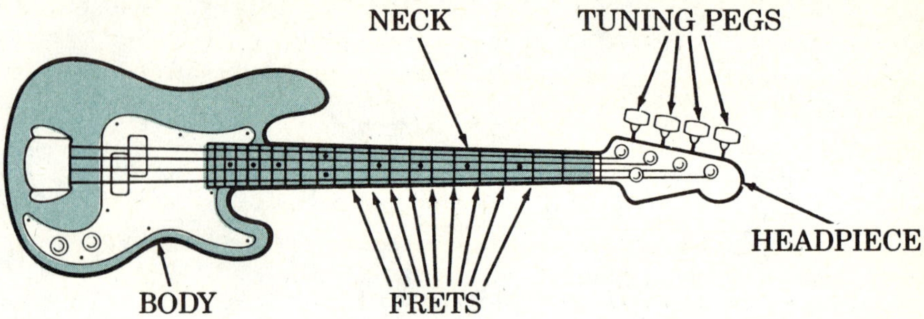

1.

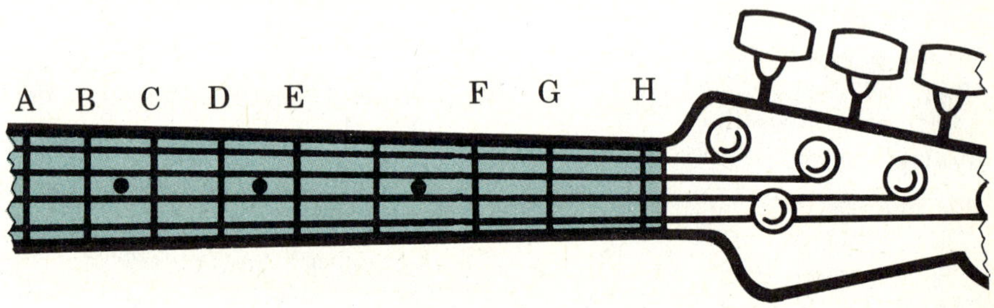

a. Fret A to Fret B __________

b. Fret A to Fret H __________

c. Fret C to Fret G __________

d. Fret B to Fret F __________

e. Fret A to Fret E __________

2.

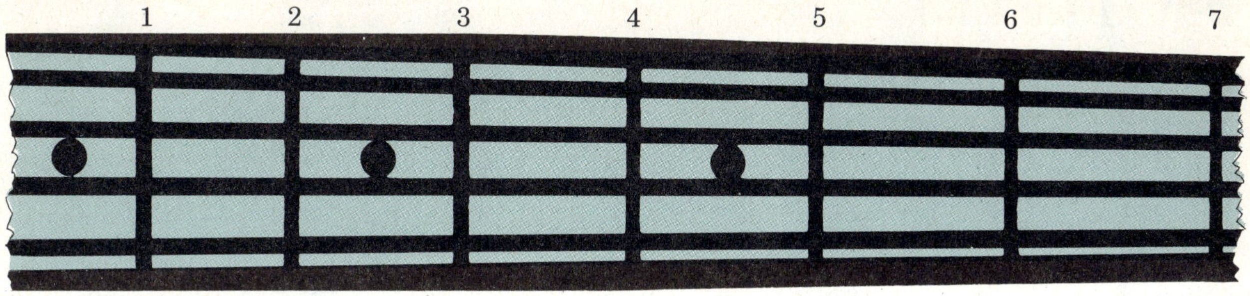

a. Fret 1 to Fret 3 ____________

b. Fret 1 to Fret 4 ____________

3.

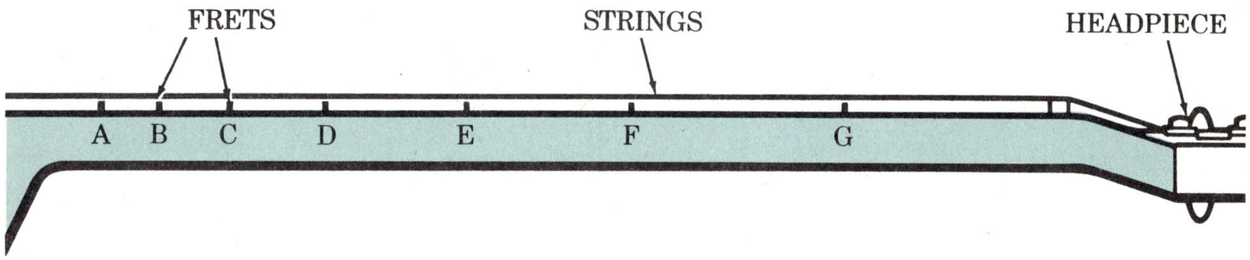

a. Fret A to Fret G ____________

b. Fret B to Fret F ____________

c. Fret B to Fret G ____________

ACTIVITY 32

DIVIDING A LINE SEGMENT INTO EQUAL PARTS

Manager________________________ Date________________

Foundations: Publicity posters must be carefully designed so they are attractive and draw attention. Letters must be evenly spaced.

Directions: Below are several rectangles that will be used on a publicity poster. Each rectangle must contain the name of a club. The letters must be evenly spaced. Divide each rectangle into the correct number of equal parts.

♪ : Do not count the apostrophe when counting the letters.

1. SHARK'S

2. CAPRICORN'S

3. GLITTER'S

4. STARZ

ACTIVITY 33

ART AND TEXT LAYOUT

Foundations: Although you are a band manager, you are not necessarily an artist. Your layout ideas are important to the artists and printers who are working on your posters and record covers.

Directions: On a blank sheet of $8\frac{1}{2}$ inch × 11 inch unlined paper, accurately construct the layout shown for the following.

♪ : Remember to mark two points to draw horizontal and vertical lines.

Manager______________________________ Date______________________

1. Prepare a layout for a concert publicity poster for your band. Use the following diagram to draw the layout; make sure you use and label the accurate measurements. Write the name of your latest album and your latest single, concert tour dates, and two sponsors in the boxes. Sponsors are corporations that contribute funds to the concerts.

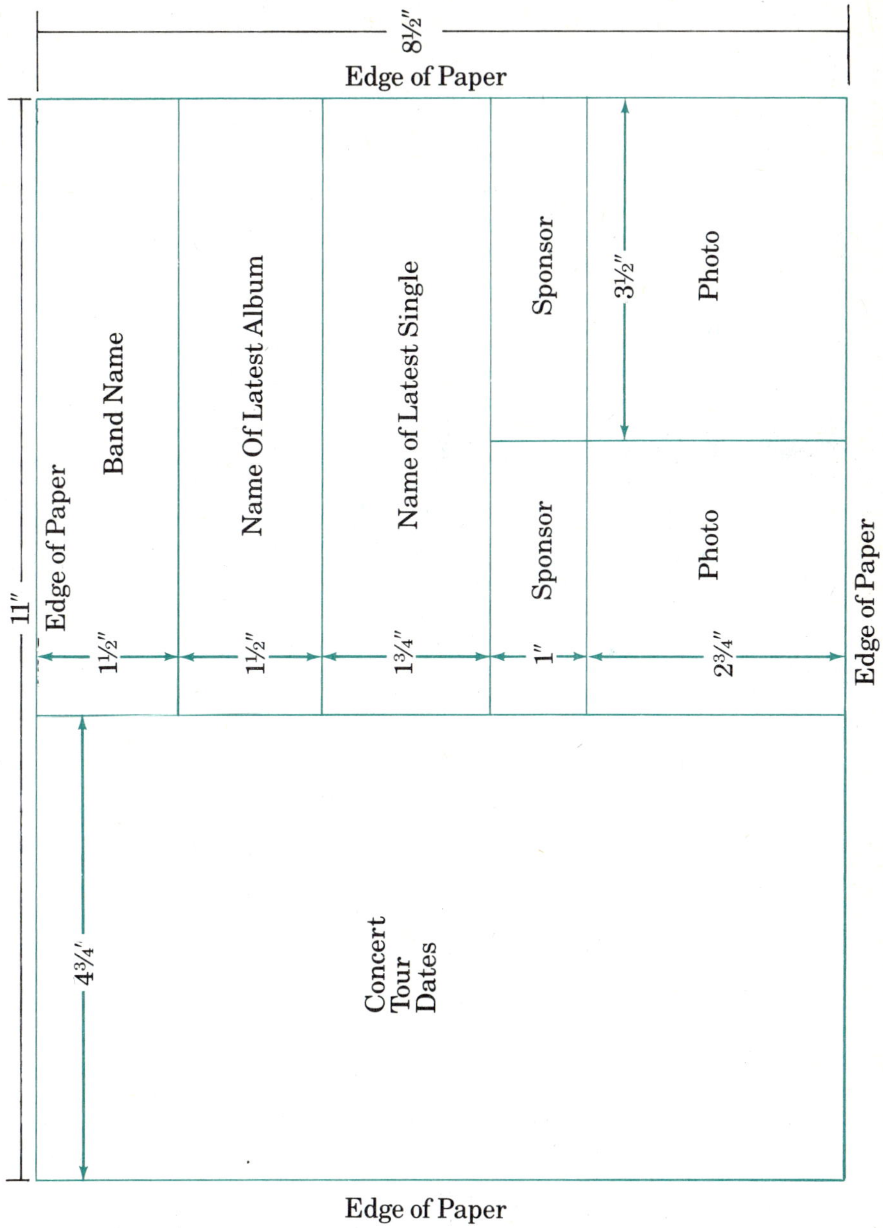

2. Prepare a layout for a publicity poster for your band. Use the following diagram to draw the layout; make sure you use and clearly label the accurate measurements. Clearly write your band's name, concert hall, and concert dates in the boxes.

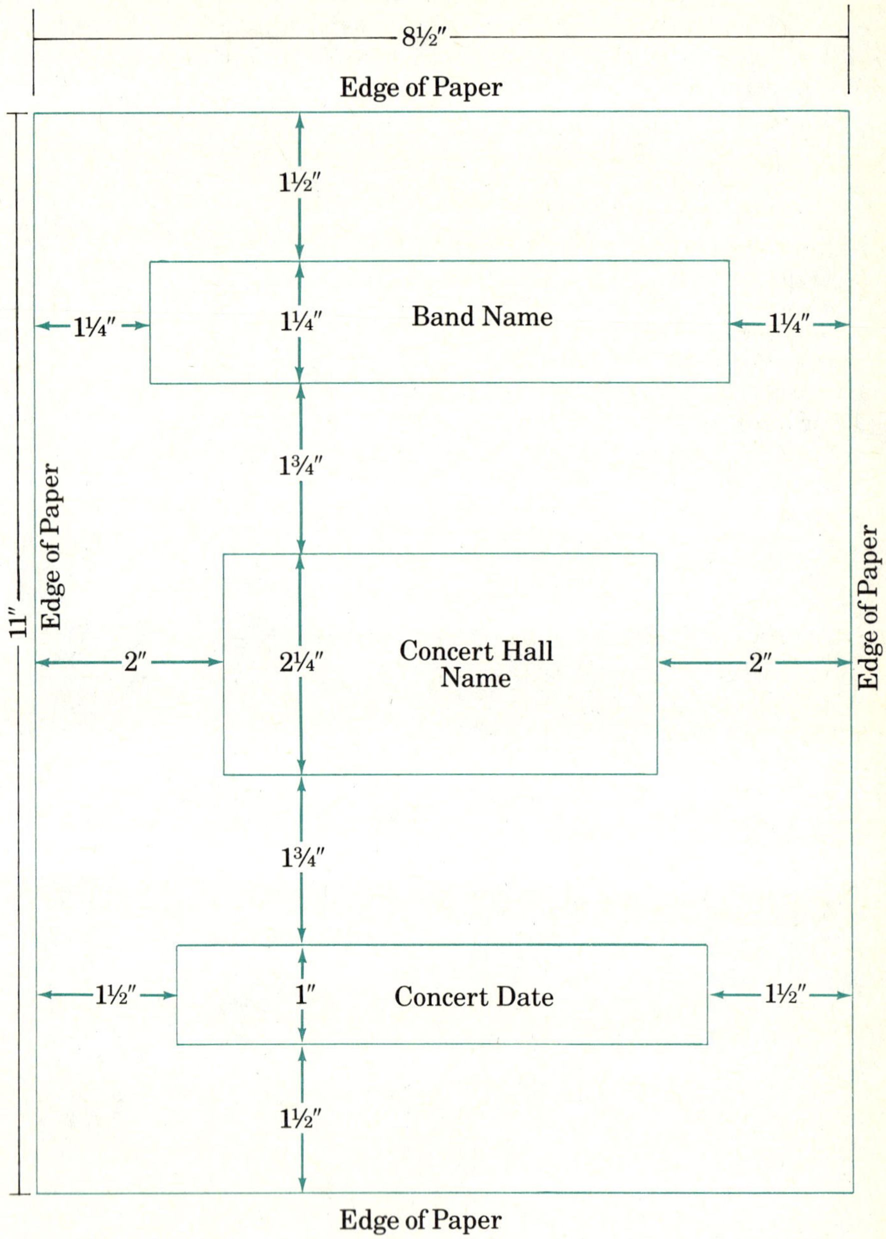

3. A 45 r.p.m. record is called a *single.* A single comes in a paper cover which is called a *sleeve.* Sleeves measure 7 inch by 7 inch. Prepare the layout for a single sleeve. Use the following diagram to draw the layout; be sure to include the name of your band and the song.

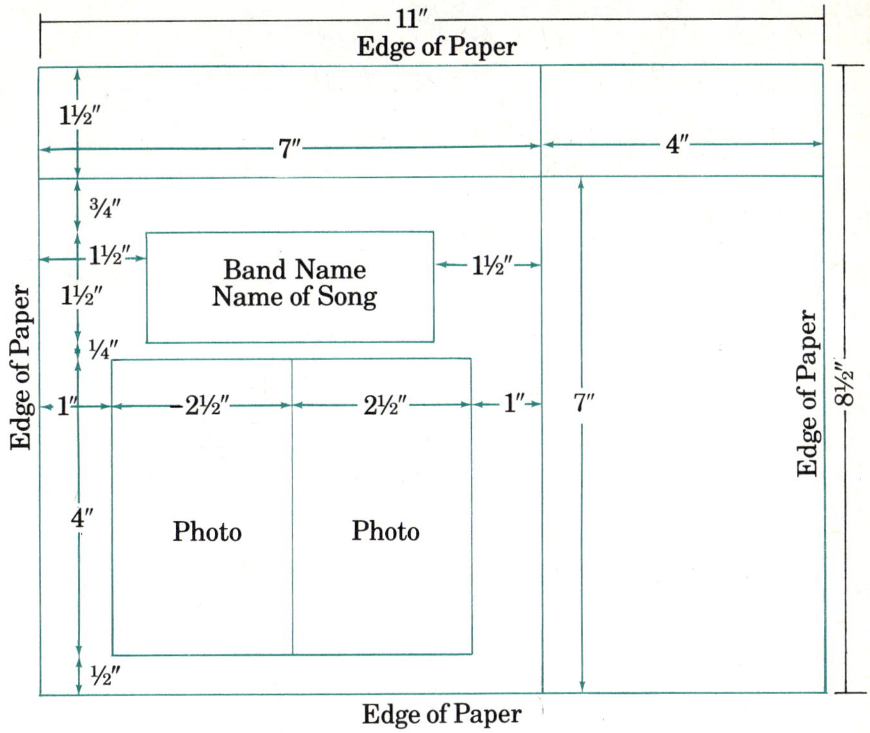

4. Prepare a layout for a publicity poster. Use the following diagram to draw the layout. Include your band's name and the measurements.

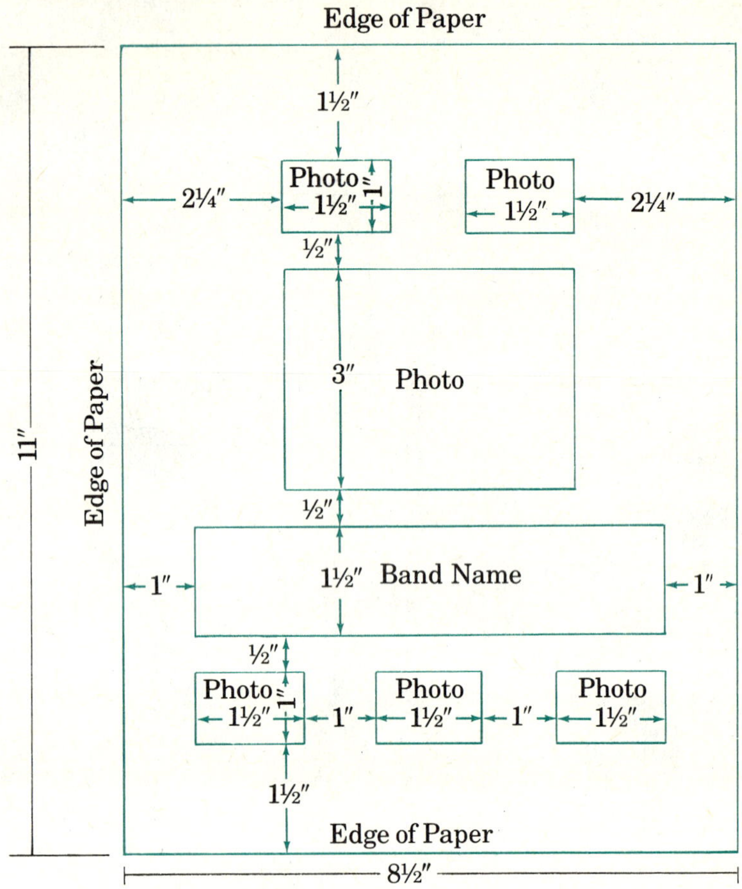

ACTIVITY 34
PUBLICITY

Manager____________________________ Date____________________

Foundations: Publicity will help increase your band's popularity. It costs money to advertise, but publicity increases your Quality Point total.

Directions: Linda Spooner has $80,000 in her Check Register. She would like to purchase two radio spots (60 seconds each), five posters, and a half-page newspaper ad. Complete the following Milestone 7/Form M – Publicity form for her publicity campaign.

P R A C T I C E V E R S I O N

MILESTONE 7
FORM M

PUBLICITY

Find your spending limit for this publicity campaign:

1. What is your Check Register balance? 1. ____________________
2. What is 6% of your Check Register balance? This is your spending limit for publicity. 2. ____________________

TYPE OF AD	UNIT PRICE	QUANTITY	QUALITY POINTS/UNIT	TOTAL PRICE	QUALITY POINTS
Poster (1 ream)	$ 450		1		
Newspaper (full page – 1 day)	$2,000		3		
Newspaper ($\frac{1}{2}$ page – 1 day)	$1,200		2		
Radio Spot (60 seconds)	$ 200		4		
Radio Spot (30 seconds)	$ 135		3		
Radio Spot (15 seconds)	$ 80		2		
Fan Club Mailing List	$ 175		3		
			TOTALS		
			$7\frac{1}{4}$% TAX		
			TOTAL COST		

Manager________________________ Date________________________

MILESTONE 7
FORM M

PUBLICITY

Find your spending limit for this publicity campaign:

1. What is your Check Register balance? 1. ____________
2. What is 6% of your Check Register balance? This is your spending limit for publicity. 2. ____________
3. After completing this form, record the payment in your Check Register.

TYPE OF AD	UNIT PRICE	QUANTITY	QUALITY POINTS/UNIT	TOTAL PRICE	QUALITY POINTS
Poster (1 ream)	$ 450		1		
Newspaper (full page – 1 day)	$2,000		3		
Newspaper ($\frac{1}{2}$ page – 1 day)	$1,200		2		
Radio Spot (60 seconds)	$ 200		4		
Radio Spot (30 seconds)	$ 135		3		
Radio Spot (15 seconds)	$ 80		2		
Fan Club Mailing List	$ 175		3		
			TOTALS		
			$7\frac{1}{4}$% TAX		
			TOTAL COST		

SEE ACTIVITY 34 FOR PRACTICE VERSION

Manager________________ Date________________

MILESTONE 7
FORM M

PUBLICITY

Find your spending limit for this publicity campaign:

1. What is your Check Register balance? 1. __________
2. What is 6% of your Check Register balance? This is your spending limit for publicity. 2. __________
3. After completing this form, record the payment in your Check Register.

TYPE OF AD	UNIT PRICE	QUANTITY	QUALITY POINTS/UNIT	TOTAL PRICE	QUALITY POINTS
Poster (1 ream)	$ 450		1		
Newspaper (full page – 1 day)	$2,000		3		
Newspaper ($\frac{1}{2}$ page – 1 day)	$1,200		2		
Radio Spot (60 seconds)	$ 200		4		
Radio Spot (30 seconds)	$ 135		3		
Radio Spot (15 seconds)	$ 80		2		
Fan Club Mailing List	$ 175		3		
			TOTALS		
			$7\frac{1}{4}$% TAX		
			TOTAL COST		

SEE ACTIVITY 34 FOR PRACTICE VERSION

Manager____________________ Date____________________

MILESTONE 7
FORM M

PUBLICITY

Find your spending limit for this publicity campaign:

1. What is your Check Register balance? 1. ____________
2. What is 6% of your Check Register balance? This is your spending limit for publicity. 2. ____________
3. After completing this form, record the payment in your Check Register.

TYPE OF AD	UNIT PRICE	QUANTITY	QUALITY POINTS/UNIT	TOTAL PRICE	QUALITY POINTS
Poster (1 ream)	$ 450		1		
Newspaper (full page – 1 day)	$2,000		3		
Newspaper ($\frac{1}{2}$ page – 1 day)	$1,200		2		
Radio Spot (60 seconds)	$ 200		4		
Radio Spot (30 seconds)	$ 135		3		
Radio Spot (15 seconds)	$ 80		2		
Fan Club Mailing List	$ 175		3		
			TOTALS		
			$7\frac{1}{4}$% TAX		
			TOTAL COST		

SEE ACTIVITY 34 FOR PRACTICE VERSION

Manager____________________ Date____________________

MILESTONE 7
FORM M

PUBLICITY

Find your spending limit for this publicity campaign:

1. What is your Check Register balance? 1. ____________
2. What is 6% of your Check Register balance? This is your spending limit for publicity. 2. ____________
3. After completing this form, record the payment in your Check Register.

TYPE OF AD	UNIT PRICE	QUANTITY	QUALITY POINTS/UNIT	TOTAL PRICE	QUALITY POINTS
Poster (1 ream)	$ 450		1		
Newspaper (full page – 1 day)	$2,000		3		
Newspaper ($\frac{1}{2}$ page – 1 day)	$1,200		2		
Radio Spot (60 seconds)	$ 200		4		
Radio Spot (30 seconds)	$ 135		3		
Radio Spot (15 seconds)	$ 80		2		
Fan Club Mailing List	$ 175		3		
			TOTALS		
			$7\frac{1}{4}$% TAX		
			TOTAL COST		

SEE ACTIVITY 34 FOR PRACTICE VERSION

Manager________________________ Date________________________

MILESTONE 7
FORM M

PUBLICITY

Find your spending limit for this publicity campaign:

1. What is your Check Register balance? 1. ____________

2. What is 6% of your Check Register balance? This is your spending limit for publicity. 2. ____________

3. After completing this form, record the payment in your Check Register.

TYPE OF AD	UNIT PRICE	QUANTITY	QUALITY POINTS/UNIT	TOTAL PRICE	QUALITY POINTS
Poster (1 ream)	$ 450		1		
Newspaper (full page – 1 day)	$2,000		3		
Newspaper ($\frac{1}{2}$ page – 1 day)	$1,200		2		
Radio Spot (60 seconds)	$ 200		4		
Radio Spot (30 seconds)	$ 135		3		
Radio Spot (15 seconds)	$ 80		2		
Fan Club Mailing List	$ 175		3		
			TOTALS		
			$7\frac{1}{4}$% TAX		
			TOTAL COST		

SEE ACTIVITY 34 FOR PRACTICE VERSION

Manager________________________ Date________________________

MILESTONE 7
FORM M

PUBLICITY

Find your spending limit for this publicity campaign:

1. What is your Check Register balance? 1. ____________
2. What is 6% of your Check Register balance? This is your spending limit for publicity. 2. ____________
3. After completing this form, record the payment in your Check Register.

TYPE OF AD	UNIT PRICE	QUANTITY	QUALITY POINTS/UNIT	TOTAL PRICE	QUALITY POINTS
Poster (1 ream)	$ 450		1		
Newspaper (full page – 1 day)	$2,000		3		
Newspaper ($\frac{1}{2}$ page – 1 day)	$1,200		2		
Radio Spot (60 seconds)	$ 200		4		
Radio Spot (30 seconds)	$ 135		3		
Radio Spot (15 seconds)	$ 80		2		
Fan Club Mailing List	$ 175		3		
			TOTALS		
			$7\frac{1}{4}$% TAX		
			TOTAL COST		

SEE ACTIVITY 34 FOR PRACTICE VERSION

Manager________________________ Date________________________

MILESTONE 7
FORM M

PUBLICITY

Find your spending limit for this publicity campaign:

1. What is your Check Register balance? 1. ____________
2. What is 6% of your Check Register balance? This is your spending limit for publicity. 2. ____________
3. After completing this form, record the payment in your Check Register.

TYPE OF AD	UNIT PRICE	QUANTITY	QUALITY POINTS/UNIT	TOTAL PRICE	QUALITY POINTS
Poster (1 ream)	$ 450		1		
Newspaper (full page – 1 day)	$2,000		3		
Newspaper ($\frac{1}{2}$ page – 1 day)	$1,200		2		
Radio Spot (60 seconds)	$ 200		4		
Radio Spot (30 seconds)	$ 135		3		
Radio Spot (15 seconds)	$ 80		2		
Fan Club Mailing List	$ 175		3		
			TOTALS		
			$7\frac{1}{4}$% TAX		
			TOTAL COST		

SEE ACTIVITY 34 FOR PRACTICE VERSION

Manager____________________ Date____________________

MILESTONE 7
FORM M

PUBLICITY

Find your spending limit for this publicity campaign:

1. What is your Check Register balance? 1. ________________
2. What is 6% of your Check Register balance? This is your spending limit for publicity. 2. ________________
3. After completing this form, record the payment in your Check Register.

TYPE OF AD	UNIT PRICE	QUANTITY	QUALITY POINTS/UNIT	TOTAL PRICE	QUALITY POINTS
Poster (1 ream)	$ 450		1		
Newspaper (full page – 1 day)	$2,000		3		
Newspaper ($\frac{1}{2}$ page – 1 day)	$1,200		2		
Radio Spot (60 seconds)	$ 200		4		
Radio Spot (30 seconds)	$ 135		3		
Radio Spot (15 seconds)	$ 80		2		
Fan Club Mailing List	$ 175		3		
			TOTALS		
			$7\frac{1}{4}$% TAX		
			TOTAL COST		

SEE ACTIVITY 34 FOR PRACTICE VERSION

MILESTONE 8

LOCAL CONCERTS

ACTIVITY 35

LOCAL CONCERTS

Foundations: Your band can earn "big money" when performing at local concerts. The attendance at these concerts will depend on your quality point total.

Directions: Glen Oaks has 600 quality points. His band is going to perform at Madison Square Garden. He will also receive $10,000 from endorsing a soft drink (Sharps and Flats). Complete the Practice Version of Milestone 8/Form N – Local Concert form for Glen Oaks.

Manager____________________________ Date______________________

PRACTICE VERSION

MILESTONE 8
FORM N

LOCAL CONCERT

CONCERT HALL	CAPACITY	BOOKING FEE*
Giants Stadium	61,000	$64,000
Shea Stadium	55,300	$60,000
Yankee Stadium	51,000	$59,000
Madison Square Garden	18,100	$29,000
Nassau Coliseum	16,200	$24,000
Radio City Music Hall	5,800	$14,000
Carnegie Hall	2,600	$ 3,000
Town Hall	1,900	$ 1,500

*The booking fees in SOUND FOUNDATIONS are fictitious.

1. How many quality points do you have? 1. ______________

2. **ATTENDANCE:**

   a. Pick a "Local Concert" card. Where are you performing? 2a. ______________

   b. What is the concert hall's capacity? (See the chart.) b. ______________

   c. Divide your quality points (No. 1) by 750 and round your answer to the nearest hundredth. c. ______________

   d. Find your attendance by multiplying the answer in No. 2c by the capacity (No. 2b). d. ______________

3. **TICKET PRICE:**

   Divide your quality points by 35 and round your answer to the nearest 50 cents. (This is the cost of one ticket.) How much does one ticket cost? 3. ______________

4. Each person attending your concert must buy a ticket. What is your income from ticket sales? 4. ______________

5. Find your local concert booking fee.* (See the chart.) 5. ______________

6. Subtract the booking fee* (No. 5) from your ticket sales income (No. 4) to find the local concert profit. 6. [ ]

7. Pick a "Sharps and Flats" card. Record the card information. 7. ______________

8. What is your adjusted profit? (If No. 7 is a "Sharp," add Nos. 6 and 7. If No. 7 is a "Flat," subtract No. 7 from No. 6.) Round to the nearest dollar. 8. ______________

9. The manager receives 10% of the first $5,000 of adjusted profit and 17% of the balance of the adjusted profit. Using the following Split Percentage flowchart, what is your share? 9. ______________

SPLIT PERCENTAGE FLOWCHART

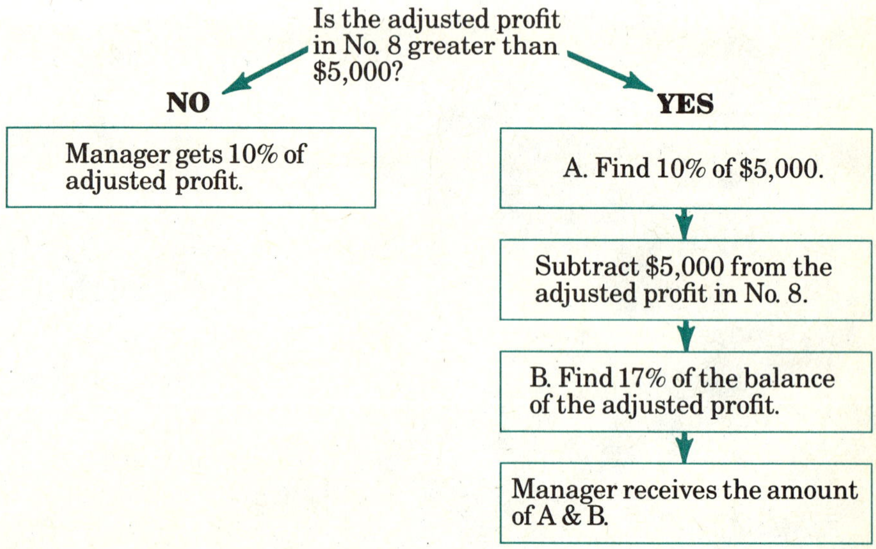

Manager____________________ Date____________________

MILESTONE 8
FORM N

LOCAL CONCERT

CONCERT HALL	CAPACITY	BOOKING FEE*
Giants Stadium	61,000	$64,000
Shea Stadium	55,300	$60,000
Yankee Stadium	51,000	$59,000
Madison Square Garden	18,100	$29,000
Nassau Coliseum	16,200	$24,000
Radio City Music Hall	5,800	$14,000
Carnegie Hall	2,600	$ 3,000
Town Hall	1,900	$ 1,500

*The booking fees in SOUND FOUNDATIONS are fictitious.

1. How many quality points do you have? 1. __________

2. **ATTENDANCE:**

   a. Pick a "Local Concert" card. Where are you performing? 2a. __________

   b. What is the concert hall's capacity? (See the chart.) b. __________

   c. Divide your quality points (No. 1) by 750 and round your answer to the nearest hundredth. c. __________

   d. Find your attendance by multiplying the answer in No. 2c by the capacity (No. 2b). d. __________

3. **TICKET PRICE:**

   Divide your quality points by 35 and round your answer to the nearest 50 cents. (This is the cost of one ticket.) How much does one ticket cost? 3. __________

4. Each person attending your concert must buy a ticket. What is your income from ticket sales? 4. __________

5. Find your local concert booking fee.* (See the chart.) 5. __________

6. Subtract the booking fee* (No. 5) from your ticket sales income (No. 4) to find your local concert profit. 6. __________

7. Pick a "Sharps and Flats" card. Record the card information. 7. ____________

8. What is your adjusted profit? (If No. 7 is a "Sharp," add Nos. 6 and 7. If No. 7 is a "Flat," subtract No. 7 from No. 6.) Round to the nearest dollar. 8. ____________

9. The manager receives 10% of the first $5,000 of adjusted profit and 17% of the balance of the adjusted profit. Using the following Split Percentage flowchart, what is your share? 9. ____________

10. Deposit the manager's share in your Check Register.

SEE ACTIVITY 35 FOR PRACTICE VERSION

SPLIT PERCENTAGE FLOWCHART

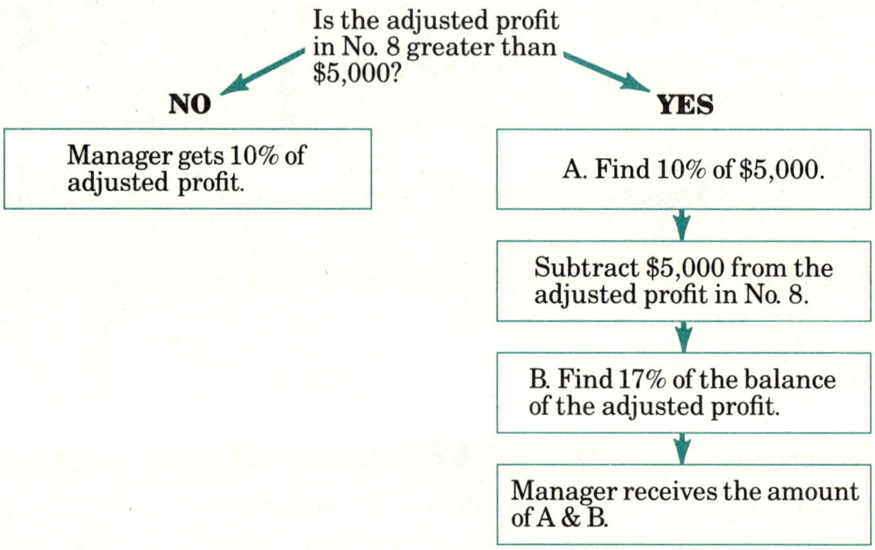

Manager____________________ Date____________________

MILESTONE 8
FORM N

LOCAL CONCERT

CONCERT HALL	CAPACITY	BOOKING FEE*
Giants Stadium	61,000	$64,000
Shea Stadium	55,300	$60,000
Yankee Stadium	51,000	$59,000
Madison Square Garden	18,100	$29,000
Nassau Coliseum	16,200	$24,000
Radio City Music Hall	5,800	$14,000
Carnegie Hall	2,600	$ 3,000
Town Hall	1,900	$ 1,500

1. How many quality points do you have? 1. ____________

2. **ATTENDANCE:**

   a. Pick a "Local Concert" card. Where are you performing? 2a. ____________

   b. What is the concert hall's capacity? (See the chart.) b. ____________

   c. Divide your quality points (No. 1) by 750 and round your answer to the nearest hundredth. c. ____________

   d. Find your attendance by multiplying the answer in No. 2c by the capacity (No. 2b). d. ____________

3. **TICKET PRICE:**

   Divide your quality points by 35 and round your answer to the nearest 50 cents. (This is the cost of one ticket.) How much does one ticket cost? 3. ____________

4. Each person attending your concert must buy a ticket. What is your income from ticket sales? 4. ____________

5. Find your local concert booking fee.* (See the chart.) 5. ____________

6. Subtract the booking fee* (No. 5) from your ticket sales income (No. 4) to find your local concert profit. 6. ____________

7. Pick a "Sharps and Flats" card. Record the card information. 7. ____________

8. What is your adjusted profit? (If No. 7 is a "Sharp," add Nos. 6 and 7. If No. 7 is a "Flat," subtract No. 7 from No. 6.) Round to the nearest dollar. 8. ____________

9. The manager receives 10% of the first $5,000 of adjusted profit and 17% of the balance of the adjusted profit. Using the following Split Percentage flowchart, what is your share? 9. ____________

10. Deposit the manager's share in your Check Register.

SEE ACTIVITY 35 FOR PRACTICE VERSION

SPLIT PERCENTAGE FLOWCHART

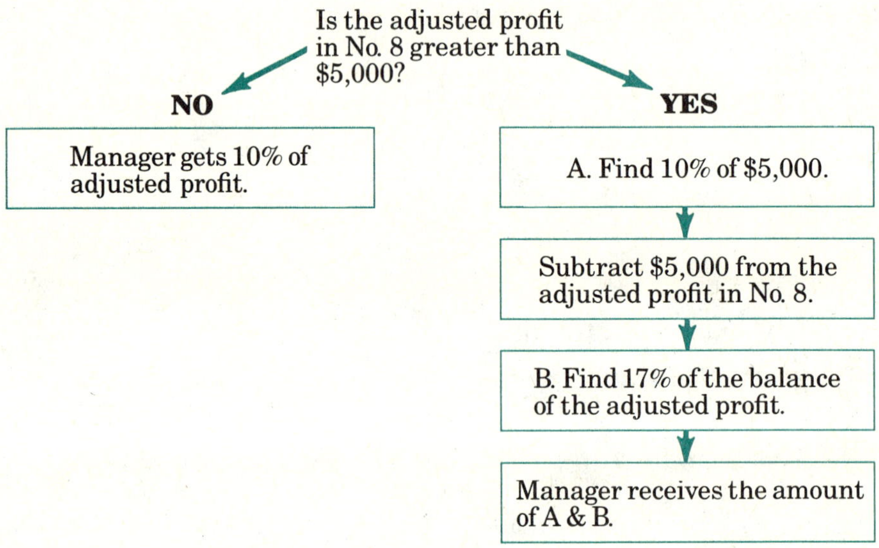

Manager________________________ Date________________________

MILESTONE 8
FORM N

LOCAL CONCERT

CONCERT HALL	CAPACITY	BOOKING FEE*
Giants Stadium	61,000	$64,000
Shea Stadium	55,300	$60,000
Yankee Stadium	51,000	$59,000
Madison Square Garden	18,100	$29,000
Nassau Coliseum	16,200	$24,000
Radio City Music Hall	5,800	$14,000
Carnegie Hall	2,600	$ 3,000
Town Hall	1,900	$ 1,500

*The booking fees in SOUND FOUNDATIONS are fictitious.

1. How many quality points do you have? 1. ____________

2. **ATTENDANCE:**

   a. Pick a "Local Concert" card. Where are you performing? 2a. ____________

   b. What is the concert hall's capacity? (See the chart.) b. ____________

   c. Divide your quality points (No. 1) by 750 and round your answer to the nearest hundredth. c. ____________

   d. Find your attendance by multiplying the answer in No. 2c by the capacity (No. 2b). d. ____________

3. **TICKET PRICE:**

   Divide your quality points by 35 and round your answer to the nearest 50 cents. (This is the cost of one ticket.) How much does one ticket cost? 3. ____________

4. Each person attending your concert must buy a ticket. What is your income from ticket sales? 4. ____________

5. Find your local concert booking fee.* (See the chart.) 5. ____________

6. Subtract the booking fee* (No. 5) from your ticket sales income (No. 4) to find your local concert profit. 6. [ ]

7. Pick a "Sharps and Flats" card. Record the card information. 7. ______________

8. What is your adjusted profit? (If No. 7 is a "Sharp," add Nos. 6 and 7. If No. 7 is a "Flat," subtract No. 7 from No. 6.) Round to the nearest dollar. 8. ______________

9. The manager receives 10% of the first $5,000 of adjusted profit and 17% of the balance of the adjusted profit. Using the following Split Percentage flowchart, what is your share? 9. ______________

10. Deposit the manager's share in your Check Register.

SEE ACTIVITY 35 FOR PRACTICE VERSION

SPLIT PERCENTAGE FLOWCHART

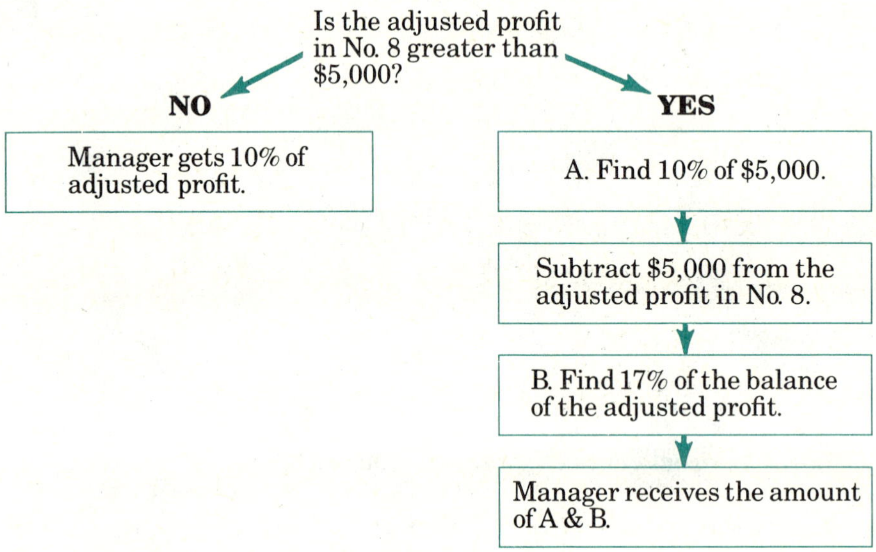

Manager____________________ Date____________________

MILESTONE 8
FORM N

LOCAL CONCERT

CONCERT HALL	CAPACITY	BOOKING FEE*
Giants Stadium	61,000	$64,000
Shea Stadium	55,300	$60,000
Yankee Stadium	51,000	$59,000
Madison Square Garden	18,100	$29,000
Nassau Coliseum	16,200	$24,000
Radio City Music Hall	5,800	$14,000
Carnegie Hall	2,600	$ 3,000
Town Hall	1,900	$ 1,500

*The booking fees in SOUND FOUNDATIONS are fictitious.

1. How many quality points do you have? 1. __________

2. **ATTENDANCE:**

   a. Pick a "Local Concert" card. Where are you performing? 2a. __________

   b. What is the concert hall's capacity? (See the chart.) b. __________

   c. Divide your quality points (No. 1) by 750 and round your answer to the nearest hundredth. c. __________

   d. Find your attendance by multiplying the answer in No. 2c by the capacity (No. 2b). d. __________

3. **TICKET PRICE:**

   Divide your quality points by 35 and round your answer to the nearest 50 cents. (This is the cost of one ticket.) How much does one ticket cost? 3. __________

4. Each person attending your concert must buy a ticket. What is your income from ticket sales? 4. __________

5. Find your local concert booking fee.* (See the chart.) 5. __________

6. Subtract the booking fee* (No. 5) from your ticket sales income (No. 4) to find your local concert profit. 6. __________

7. Pick a "Sharps and Flats" card. Record the card information. 7. ____________

8. What is your adjusted profit? (If No. 7 is a "Sharp," add Nos. 6 and 7. If No. 7 is a "Flat," subtract No. 7 from No. 6.) Round to the nearest dollar. 8. ____________

9. The manager receives 10% of the first $5,000 of adjusted profit and 17% of the balance of the adjusted profit. Using the following Split Percentage flowchart, what is your share? 9. ____________

10. Deposit the manager's share in your Check Register.

SEE ACTIVITY 35 FOR PRACTICE VERSION

SPLIT PERCENTAGE FLOWCHART

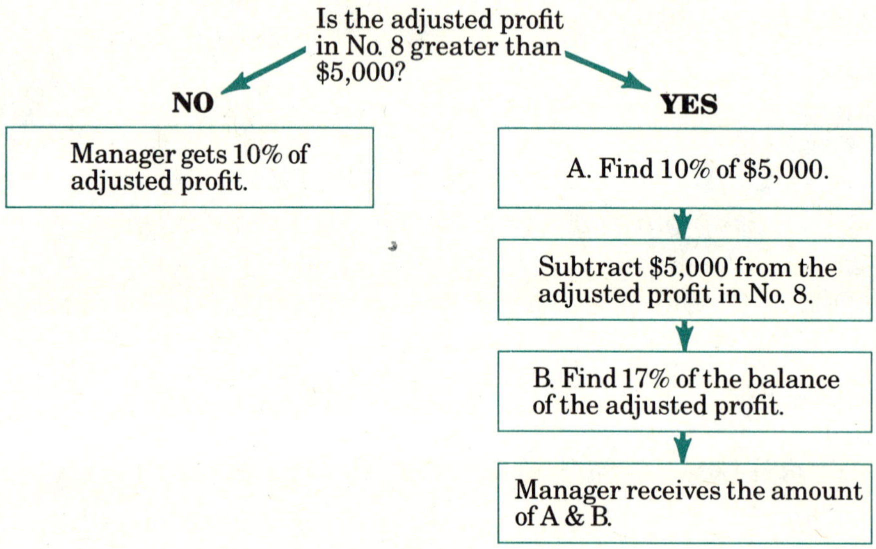

Manager________________________ Date________________________

MILESTONE 8
FORM N

LOCAL CONCERT

CONCERT HALL	CAPACITY	BOOKING FEE*
Giants Stadium	61,000	$64,000
Shea Stadium	55,300	$60,000
Yankee Stadium	51,000	$59,000
Madison Square Garden	18,100	$29,000
Nassau Coliseum	16,200	$24,000
Radio City Music Hall	5,800	$14,000
Carnegie Hall	2,600	$ 3,000
Town Hall	1,900	$ 1,500

*The booking fees in SOUND FOUNDATIONS are fictitious.

1. How many quality points do you have? 1. ____________

2. **ATTENDANCE:**

   a. Pick a "Local Concert" card. Where are you performing? 2a. ____________

   b. What is the concert hall's capacity? (See the chart.) b. ____________

   c. Divide your quality points (No. 1) by 750 and round your answer to the nearest hundredth. c. ____________

   d. Find your attendance by multiplying the answer in No. 2c by the capacity (No. 2b). d. ____________

3. **TICKET PRICE:**

   Divide your quality points by 35 and round your answer to the nearest 50 cents. (This is the cost of one ticket.) How much does one ticket cost? 3. ____________

4. Each person attending your concert must buy a ticket. What is your income from ticket sales? 4. ____________

5. Find your local concert booking fee.* (See the chart.) 5. ____________

6. Subtract the booking fee* (No. 5) from your ticket sales income (No. 4) to find your local concert profit. 6. ____________

7. Pick a "Sharps and Flats" card. Record the card information. 7. ________

8. What is your adjusted profit? (If No. 7 is a "Sharp," add Nos. 6 and 7. If No. 7 is a "Flat," subtract No. 7 from No. 6.) Round to the nearest dollar. 8. ________

9. The manager receives 10% of the first $5,000 of adjusted profit and 17% of the balance of the adjusted profit. Using the following Split Percentage flowchart, what is your share? 9. ________

10. Deposit the manager's share in your Check Register.

SEE ACTIVITY 35 FOR PRACTICE VERSION

SPLIT PERCENTAGE FLOWCHART

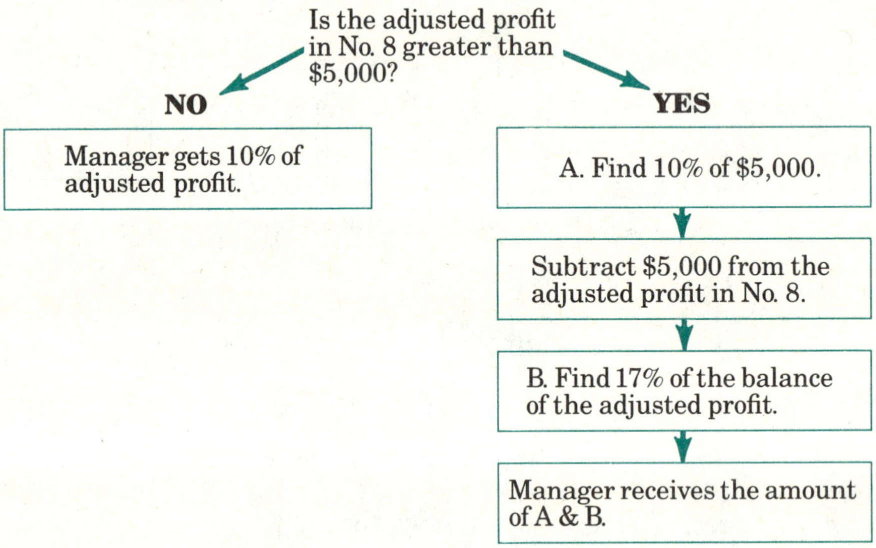

Manager__________________________ Date________________________

MILESTONE 8
FORM N

LOCAL CONCERT

CONCERT HALL	CAPACITY	BOOKING FEE*
Giants Stadium	61,000	$64,000
Shea Stadium	55,300	$60,000
Yankee Stadium	51,000	$59,000
Madison Square Garden	18,100	$29,000
Nassau Coliseum	16,200	$24,000
Radio City Music Hall	5,800	$14,000
Carnegie Hall	2,600	$ 3,000
Town Hall	1,900	$ 1,500

*The booking fees in SOUND FOUNDATIONS are fictitious.

1. How many quality points do you have? 1. __________

2. **ATTENDANCE:**

   a. Pick a "Local Concert" card. Where are you performing? 2a. __________

   b. What is the concert hall's capacity? (See the chart.) b. __________

   c. Divide your quality points (No. 1) by 750 and round your answer to the nearest hundredth. c. __________

   d. Find your attendance by multiplying the answer in No. 2c by the capacity (No. 2b). d. __________

3. **TICKET PRICE:**

   Divide your quality points by 35 and round your answer to the nearest 50 cents. (This is the cost of one ticket.) How much does one ticket cost? 3. __________

4. Each person attending your concert must buy a ticket. What is your income from ticket sales? 4. __________

5. Find your local concert booking fee.* (See the chart.) 5. __________

6. Subtract the booking fee* (No. 5) from your ticket sales income (No. 4) to find your local concert profit. 6. __________

7. Pick a "Sharps and Flats" card. Record the card information. 7. ______________

8. What is your adjusted profit? (If No. 7 is a "Sharp," add Nos. 6 and 7. If No. 7 is a "Flat," subtract No. 7 from No. 6.) Round to the nearest dollar. 8. ______________

9. The manager receives 10% of the first $5,000 of adjusted profit and 17% of the balance of the adjusted profit. Using the following Split Percentage flowchart, what is your share? 9. ______________

10. Deposit the manager's share in your Check Register.

SEE ACTIVITY 35 FOR PRACTICE VERSION

SPLIT PERCENTAGE FLOWCHART

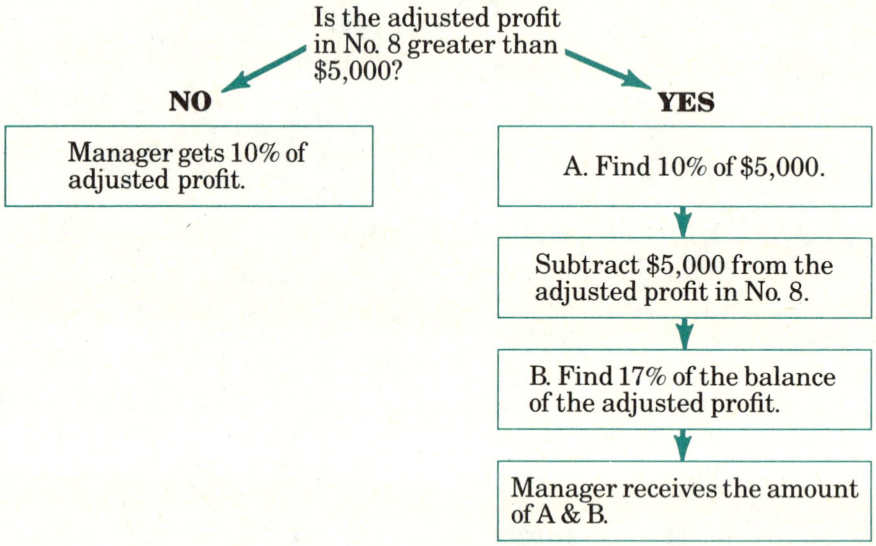

Manager________________________ Date________________________

MILESTONE 8
FORM N

LOCAL CONCERT

CONCERT HALL	CAPACITY	BOOKING FEE*
Giants Stadium	61,000	$64,000
Shea Stadium	55,300	$60,000
Yankee Stadium	51,000	$59,000
Madison Square Garden	18,100	$29,000
Nassau Coliseum	16,200	$24,000
Radio City Music Hall	5,800	$14,000
Carnegie Hall	2,600	$ 3,000
Town Hall	1,900	$ 1,500

*The booking fees in SOUND FOUNDATIONS are fictitious.

1. How many quality points do you have? 1. ____________

2. **ATTENDANCE:**

   a. Pick a "Local Concert" card. Where are you performing? 2a. ____________

   b. What is the concert hall's capacity? (See the chart.) b. ____________

   c. Divide your quality points (No. 1) by 750 and round your answer to the nearest hundredth. c. ____________

   d. Find your attendance by multiplying the answer in No. 2c by the capacity (No. 2b). d. ____________

3. **TICKET PRICE:**

   Divide your quality points by 35 and round your answer to the nearest 50 cents. (This is the cost of one ticket.) How much does one ticket cost? 3. ____________

4. Each person attending your concert must buy a ticket. What is your income from ticket sales? 4. ____________

5. Find your local concert booking fee.* (See the chart.) 5. ____________

6. Subtract the booking fee* (No. 5) from your ticket sales income (No. 4) to find your local concert profit. 6. ____________

7. Pick a "Sharps and Flats" card. Record the card information. 7. ____________

8. What is your adjusted profit? (If No. 7 is a "Sharp," add Nos. 6 and 7. If No. 7 is a "Flat," subtract No. 7 from No. 6.) Round to the nearest dollar. 8. ____________

9. The manager receives 10% of the first $5,000 of adjusted profit and 17% of the balance of the adjusted profit. Using the following Split Percentage flowchart, what is your share? 9. ____________

10. Deposit the manager's share in your Check Register.

SEE ACTIVITY 35 FOR PRACTICE VERSION

SPLIT PERCENTAGE FLOWCHART

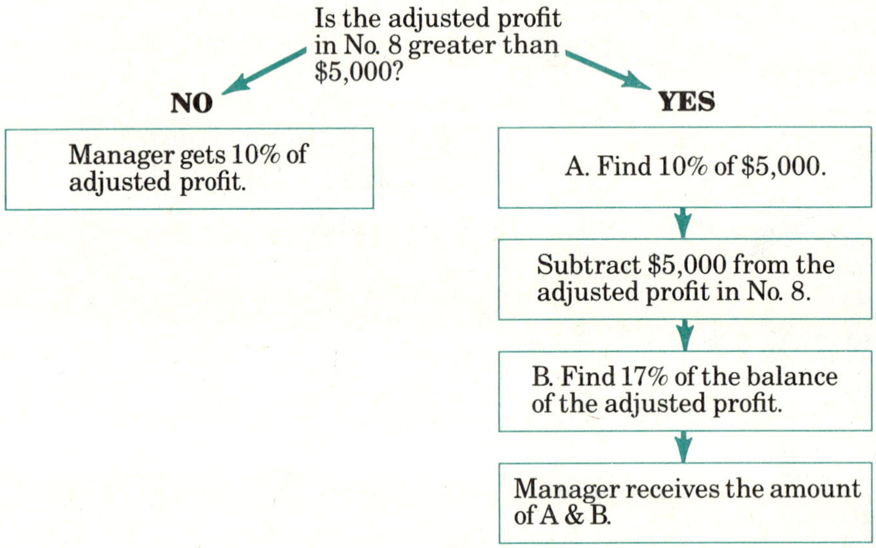

Manager____________________ Date____________________

MILESTONE 8
FORM N

LOCAL CONCERT

CONCERT HALL	CAPACITY	BOOKING FEE*
Giants Stadium	61,000	$64,000
Shea Stadium	55,300	$60,000
Yankee Stadium	51,000	$59,000
Madison Square Garden	18,100	$29,000
Nassau Coliseum	16,200	$24,000
Radio City Music Hall	5,800	$14,000
Carnegie Hall	2,600	$ 3,000
Town Hall	1,900	$ 1,500

*The booking fees in SOUND FOUNDATIONS are fictitious.

1. How many quality points do you have? 1. __________

2. **ATTENDANCE:**

   a. Pick a "Local Concert" card. Where are you performing? 2a. __________

   b. What is the concert hall's capacity? (See the chart.) b. __________

   c. Divide your quality points (No. 1) by 750 and round your answer to the nearest hundredth. c. __________

   d. Find your attendance by multiplying the answer in No. 2c by the capacity (No. 2b). d. __________

3. **TICKET PRICE:**

   Divide your quality points by 35 and round your answer to the nearest 50 cents. (This is the cost of one ticket.) How much does one ticket cost? 3. __________

4. Each person attending your concert must buy a ticket. What is your income from ticket sales? 4. __________

5. Find your local concert booking fee.* (See the chart.) 5. __________

6. Subtract the booking fee* (No. 5) from your ticket sales income (No. 4) to find your local concert profit. 6. __________

7. Pick a "Sharps and Flats" card. Record the card information. 7. ________________

8. What is your adjusted profit? (If No. 7 is a "Sharp," add Nos. 6 and 7. If No. 7 is a "Flat," subtract No. 7 from No. 6.) Round to the nearest dollar. 8. ________________

9. The manager receives 10% of the first $5,000 of adjusted profit and 17% of the balance of the adjusted profit. Using the following Split Percentage flowchart, what is your share? 9. ________________

10. Deposit the manager's share in your Check Register.

SEE ACTIVITY 35 FOR PRACTICE VERSION

SPLIT PERCENTAGE FLOWCHART

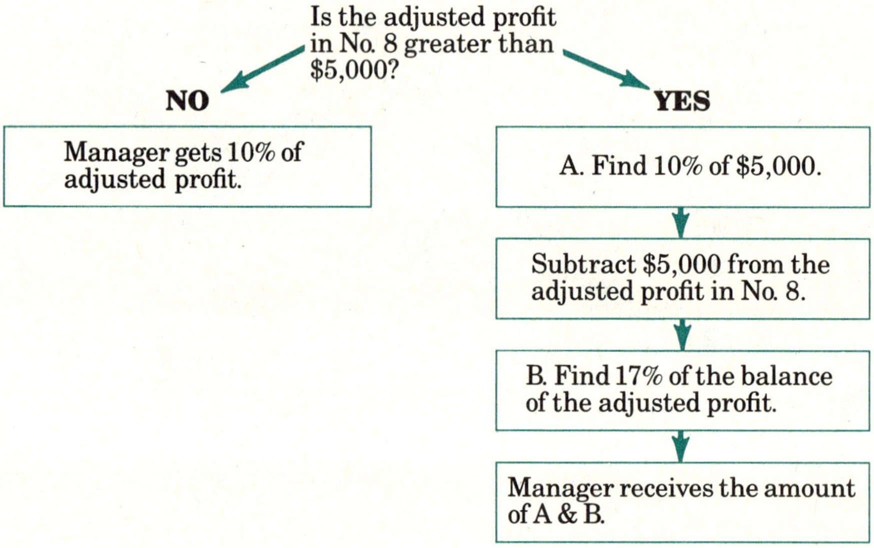

Manager______________________________ Date__________________________

MILESTONE 8
FORM N

LOCAL CONCERT

CONCERT HALL	CAPACITY	BOOKING FEE*
Giants Stadium	61,000	$64,000
Shea Stadium	55,300	$60,000
Yankee Stadium	51,000	$59,000
Madison Square Garden	18,100	$29,000
Nassau Coliseum	16,200	$24,000
Radio City Music Hall	5,800	$14,000
Carnegie Hall	2,600	$ 3,000
Town Hall	1,900	$ 1,500

*The booking fees in SOUND FOUNDATIONS are fictitious.

1. How many quality points do you have? 1. ____________

2. **ATTENDANCE:**

   a. Pick a "Local Concert" card. Where are you performing? 2a. ____________

   b. What is the concert hall's capacity? (See the chart.) b. ____________

   c. Divide your quality points (No. 1) by 750 and round your answer to the nearest hundredth. c. ____________

   d. Find your attendance by multiplying the answer in No. 2c by the capacity (No. 2b). d. ____________

3. **TICKET PRICE:**

   Divide your quality points by 35 and round your answer to the nearest 50 cents. (This is the cost of one ticket.) How much does one ticket cost? 3. ____________

4. Each person attending your concert must buy a ticket. What is your income from ticket sales? 4. ____________

5. Find your local concert booking fee.* (See the chart.) 5. ____________

6. Subtract the booking fee* (No. 5) from your ticket sales income (No. 4) to find your local concert profit. 6. [ ]

7. Pick a "Sharps and Flats" card. Record the card information. 7. ______________

8. What is your adjusted profit? (If No. 7 is a "Sharp," add Nos. 6 and 7. If No. 7 is a "Flat," subtract No. 7 from No. 6.) Round to the nearest dollar. 8. ______________

9. The manager receives 10% of the first $5,000 of adjusted profit and 17% of the balance of the adjusted profit. Using the following Split Percentage flowchart, what is your share? 9. ______________

10. Deposit the manager's share in your Check Register.

SEE ACTIVITY 35 FOR PRACTICE VERSION

SPLIT PERCENTAGE FLOWCHART

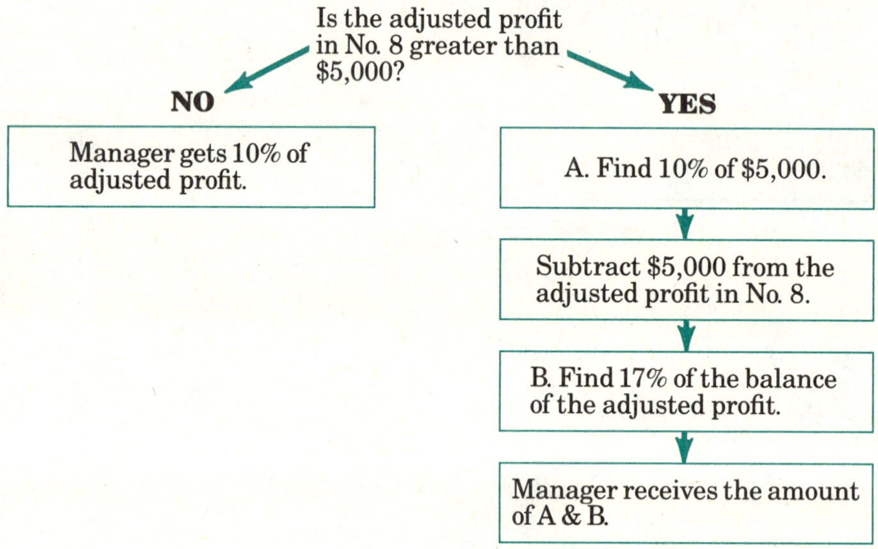

Manager________________________ Date________________________

MILESTONE 8
FORM N

LOCAL CONCERT

CONCERT HALL	CAPACITY	BOOKING FEE*
Giants Stadium	61,000	$64,000
Shea Stadium	55,300	$60,000
Yankee Stadium	51,000	$59,000
Madison Square Garden	18,100	$29,000
Nassau Coliseum	16,200	$24,000
Radio City Music Hall	5,800	$14,000
Carnegie Hall	2,600	$ 3,000
Town Hall	1,900	$ 1,500

*The booking fees in SOUND FOUNDATIONS are fictitious.

1. How many quality points do you have? 1. ____________

2. **ATTENDANCE:**

   a. Pick a "Local Concert" card. Where are you performing? 2a. ____________

   b. What is the concert hall's capacity? (See the chart.) b. ____________

   c. Divide your quality points (No. 1) by 750 and round your answer to the nearest hundredth. c. ____________

   d. Find your attendance by multiplying the answer in No. 2c by the capacity (No. 2b). d. ____________

3. **TICKET PRICE:**

   Divide your quality points by 35 and round your answer to the nearest 50 cents. (This is the cost of one ticket.) How much does one ticket cost? 3. ____________

4. Each person attending your concert must buy a ticket. What is your income from ticket sales? 4. ____________

5. Find your local concert booking fee.* (See the chart.) 5. ____________

6. Subtract the booking fee* (No. 5) from your ticket sales income (No. 4) to find your local concert profit. 6. ____________

7. Pick a "Sharps and Flats" card. Record the card information. 7. ____________

8. What is your adjusted profit? (If No. 7 is a "Sharp," add Nos. 6 and 7. If No. 7 is a "Flat," subtract No. 7 from No. 6.) Round to the nearest dollar. 8. ____________

9. The manager receives 10% of the first $5,000 of adjusted profit and 17% of the balance of the adjusted profit. Using the following Split Percentage flowchart, what is your share? 9. ____________

10. Deposit the manager's share in your Check Register.

SEE ACTIVITY 35 FOR PRACTICE VERSION

SPLIT PERCENTAGE FLOWCHART

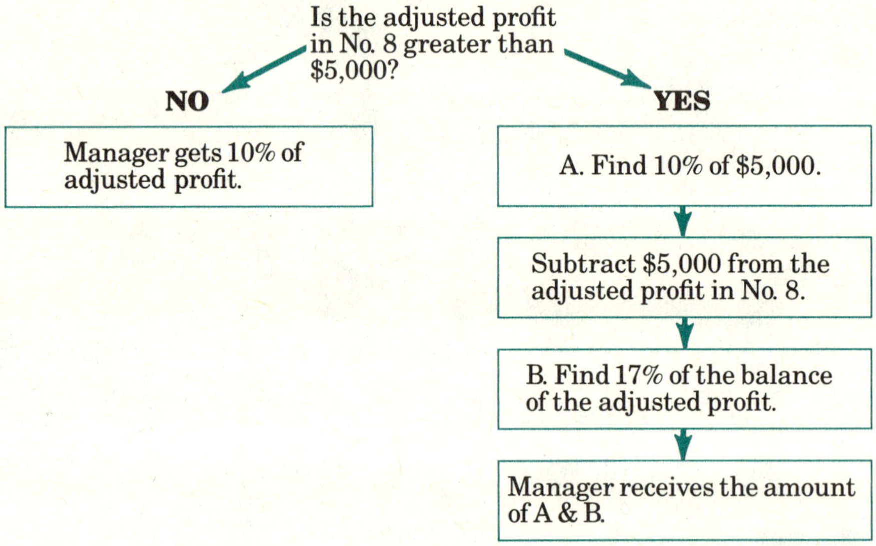

ACTIVITY 36
PERIMETER

Foundations: The light and sound crew are responsible for carefully planning stage layout, electrical wiring, and stage lighting. They must map out the paths for all cables and wires around the outside of the stage (for safety reasons). A manager must know the perimeter of each stage so that the correct amount of wire can be purchased.

Directions: Answer each of the following questions.

Manager____________________________ Date_______________________

1. Your next concert will be at Washington Square Park. This stage is rectangular. The length is 29 feet and the width is 19 feet. What is the perimeter of this stage? (Draw a sketch.)

1. ____________________

2. Your band has been chosen to perform at the Pentagon for their "Spring Fling." The floor plan of the stage is shown here. What is the perimeter of the stage?

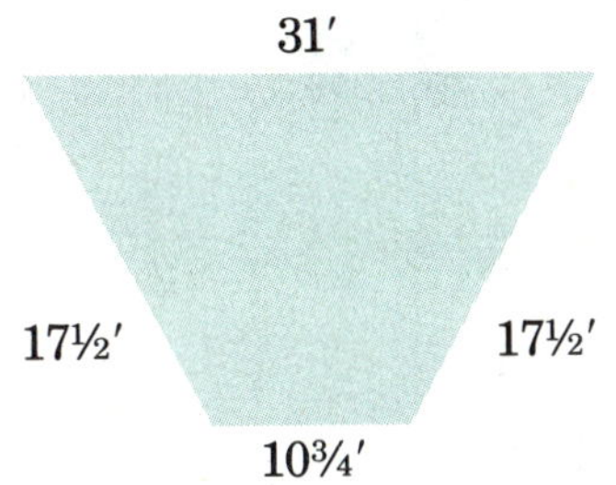

2. ____________________

3. The manager of Starz is planning to build a new, star-shaped stage. Each side is 10.5 meters. The outside edge of the stage will be decorated with a string of gold stars. How long should the string of gold stars be?

3. ____________________

4. Your roadies string disco lighting around the edge of the stage at different clubs. Use the following diagrams to find the perimeter of each stage.

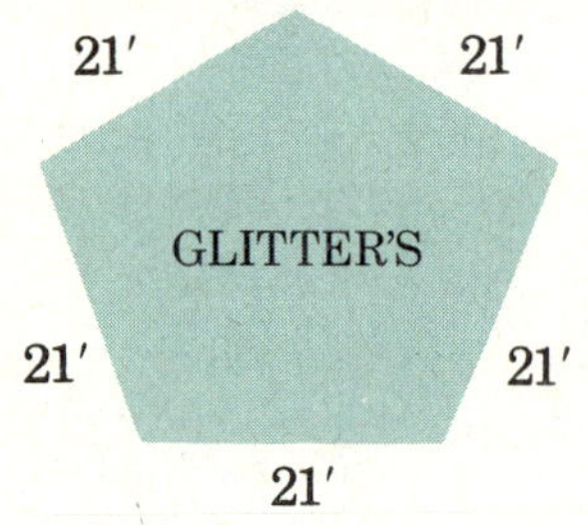

4a. The perimeter of Glitter's stage is

____

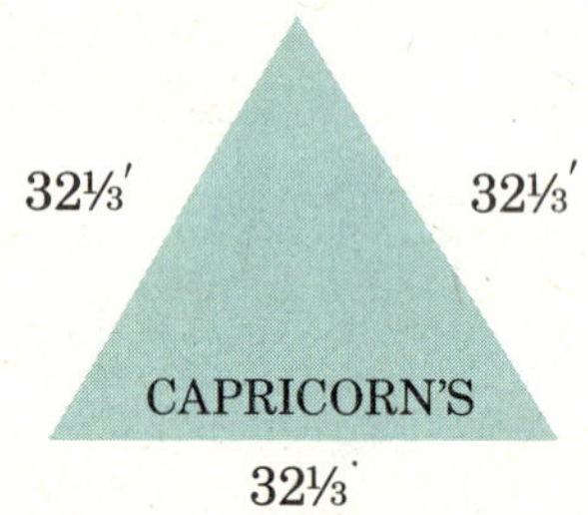

b. The perimeter of Capricorn's stage is

____

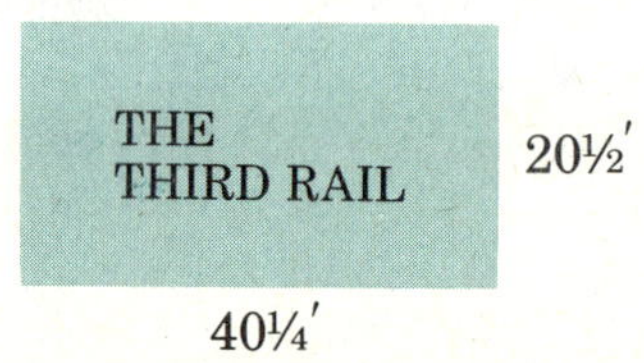

c. The perimeter of The Third Rail's stage is

____

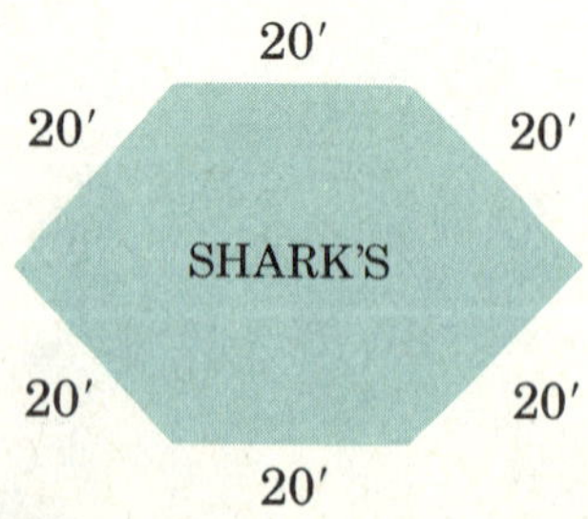

d. The perimeter of Shark's stage is

____

5. At which clubs in No. 4 would 120 feet of disco lighting be enough to cover the stage perimeter?

____

Manager________________________ Date________________

ACTIVITY 37

THE CIRCUMFERENCE OF A CIRCLE

Foundations: The drummer is usually elevated on a platform called a *drum riser.* The drum riser is usually a circular wooden platform that is supported with legs. Fabric is sometimes hung around the platform to hide the supports. You must know the circumference of the drum riser to purchase the correct length of fabric.

Directions: Using $\pi = 3.14$, answer each of the following questions.

♪ : Remember, the radius of a circle is equal to one half the diameter.

1. The diameter of a drum riser is 8 feet. How many feet of fabric are needed to completely surround the riser?

   1. ________________

2. The radius of a drum riser is 7 feet. You are going to hang decorative lights around the edge of the drum riser. What is the circumference of the drum riser?

   2. ________________

3. Your band will be playing in Canada next month. The roadies called the arena and found out that the drum riser they will have is 3.1 meters in diameter. The roadies want to purchase fancy black satin material to hang from the riser. What is the least amount of fabric that must be purchased?

   3. ________________

4. The Brooklyn Bridge has a square drum riser. Each side is 9 feet long.

   a. What is the perimeter of the drum riser? 4a. ________________

   b. The circular drum riser at Shark's has a diameter of 9 feet. What is the circumference of the drum riser at Shark's? b. ________________

   c. If you had exactly enough material to surround the drum riser at Shark's, could that piece of material also surround the drum riser at The Brooklyn Bridge? c. ________________

5. The stage at the City Zoo is shown in this diagram. You will hang a curtain along the front semicircle of the stage. How long is the semicircle?

   5. ________________

6. Fabric costs $11 per foot. Your band's drum riser is $6\frac{1}{2}$ feet in diameter. The entire edge of the drum riser must be covered with this fabric. How much will the fabric cost to cover the edge of the drum riser?

   6. ________________

Manager________________________________ Date____________________

ACTIVITY 38
AREA

Foundations: A recording studio must be built very carefully. This usually involves carpeting floors, walls, ceilings, and drum risers. In order to determine the amount of carpeting needed, you must know how to compute the area.

Directions: To the right are diagrams of the studio in which your band recorded its demo. Answer each of the following questions.

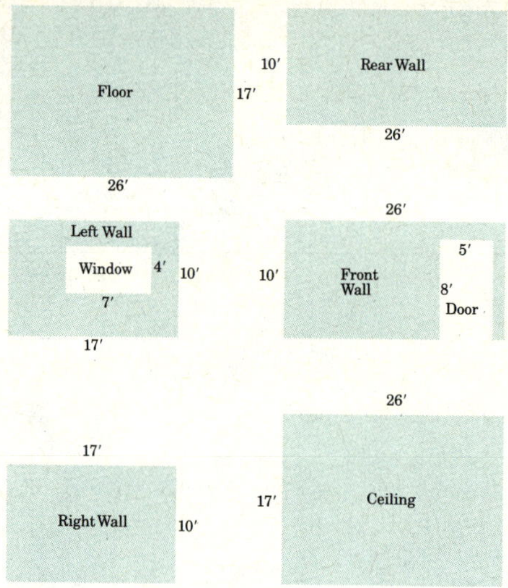

1. What is the area of the right wall?

1. ____________________

2. What is the area of the floor?

2. ____________________

3. What is the area of the ceiling?

3. ____________________

4. If carpeting costs $6.50 per square foot, how much will it cost to carpet the rear wall?

4. ____________________

5. Use the diagram of the left wall.
   a. What is the area of the wall including the window? 5a. ____________________
   b. What is the area of the window? b. ____________________
   c. How much carpet would you need to cover the left wall? c. ____________________

6. Use the diagram of the front wall.
   a. What is the area of the front wall including the door? 6a. ____________________
   b. What is the area of the door? b. ____________________
   c. How much carpet would you need to cover the front wall? c. ____________________

7. How much carpet would you need to cover a drum riser that is 10 feet in diameter?

7. ____________________

ACTIVITY 39
VOLUME

Manager____________________ Date________________

Foundations: The speaker cabinets used in home stereo systems and large P.A. systems are very important to the quality of the sound. Different cabinets will change the sound of the speaker. Electricians and musicians must know how to compute volume to manufacture speaker systems.

Directions: Find the volume of each of the following speaker cabinets.

♪ : Remember – your answer must be in cubic units!

1. Volume = __________

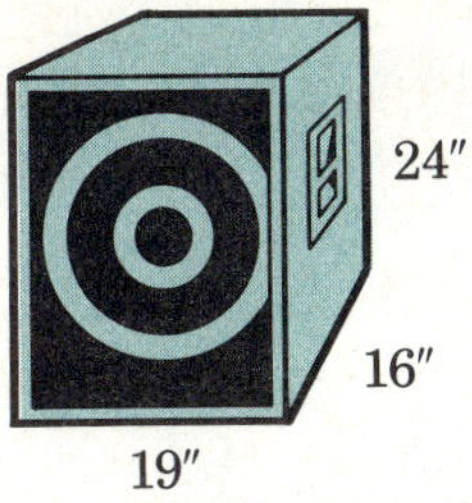

"THE STEREO KING"

2. Volume = __________

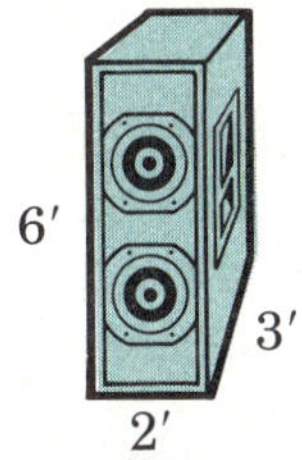

"THE TOWER"

3. Volume = __________

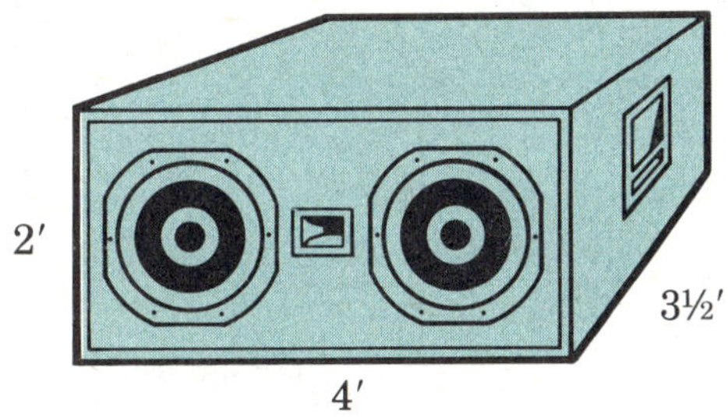

"THE BIG BAD BASS BIN"

4. Volume = __________

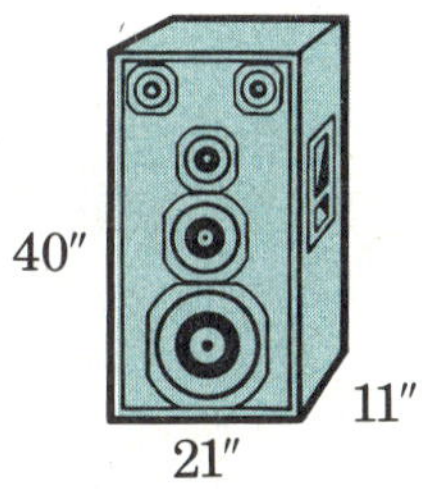

"THE THINLINE"

5. Volume = __________

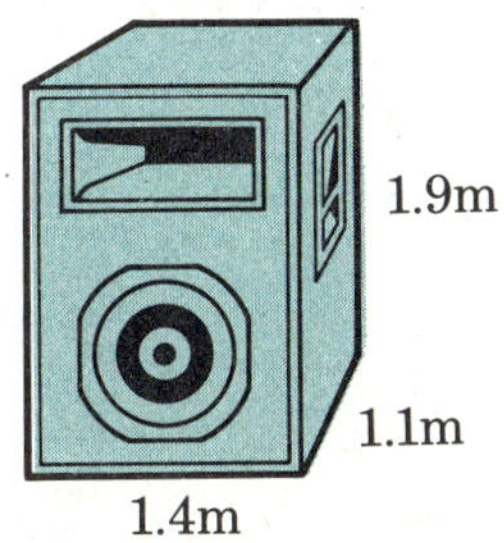

"THE COLBY CABINET"

6. The front of "The Stereo King" cabinet (No. 1) is covered with material (called *grillecloth*) to protect the speaker. What is the *area* of the grillecloth needed for "The Stereo King"?

____

MILESTONE

9

AWAY CONCERTS

ACTIVITY 40
GASOLINE EXPENSES

Manager___________________________ Date____________________

Foundations: Your band's popularity can "skyrocket" when you travel to away concerts. The band gets more exposure by traveling to perform for fans nationwide. However, make sure you have taken care of all the necessities before you "hit the road." You must be able to estimate gasoline expenses so you can set aside enough money to buy gas.

Directions: Answer the following.

1. Fill in the blanks in this chart. (Round your answers to two decimal places.)

FROM	TO	MILES TRAVELED	GALLONS USED	M.P.G.
a. Norfolk, VA	Albuquerque, NM	1,888	100	
b. Pittsburg, PA	Atlanta, GA	734		15
c. St. Louis, MO	Baltimore, MD	895	16	
d. Tampa, FL	Boston, MA	1,130		25
e. Los Angeles, CA	Miami, FL	2,769	110	

2. You must travel from Dallas, Texas, to Albuquerque, New Mexico, in order to play at a special benefit concert. The distance between these two cities is 650 miles. If your truck gets 13 miles per gallon, approximately how many gallons of gas will you need to make the trip?

2. ________________

3. Your band is in Albuquerque, New Mexico. Your friend owns a dance club in Kansas City, Missouri. He said that he would pay all your expenses if you played at his club. The distance from Albuquerque to Kansas City is 780 miles. If your truck gets 13 miles per gallon, and gas costs $1.50 per gallon, approximately how much will the gas expense be for the trip?

3. ________________

4. After a tune-up, you now find that your truck averages 18 miles per gallon. You must travel from Cincinnati, Ohio, to New Orleans, Louisiana, in order to play at a Mardi Gras party. The distance from Cincinnati to New Orleans is 792 miles. If gas costs $1.50 per gallon, what will the gas expenses be?

4. ________________

ACTIVITY 41
AWAY CONCERTS

Manager____________________ Date____________

Foundations: The band gets more exposure by traveling to perform for fans nationwide. Away concert forms are similar to local concert forms (Milestone 8/Form N) but they also include hotel expenses. The band must stay in a hotel after a concert because it is late and the members are too tired to travel long distances.

Directions: Pat Groom has 600 quality points. Her band booked an away concert in Philadelphia at the Convention Hall. But she was involved in a car accident that will cost her $737 (Sharps and Flats). Complete a Practice Version of Milestone 9/Form O – Away Concert form for Pat.

PRACTICE VERSION

MILESTONE 9
FORM O

AWAY CONCERT

AWAY CONCERT HALL	LOCATION	CAPACITY	DISTANCE FROM NEW YORK CITY	BOOKING FEE*	STATE TAX RATE
Atlantic City Auditorium	Atlantic City, NJ	40,000	150 miles	$19,000	6%
Philadelphia Convention Hall	Philadelphia, PA	9,500	100 miles	$ 9,000	6%
Veteran's Coliseum	New Haven, CT	8,700	80 miles	$ 8,000	7.5%
Hershey Park Arena	Hershey, PA	7,200	160 miles	$ 7,000	6%
J. F. K. Center	Washington, DC	2,700	230 miles	$ 2,100	6%

*The booking fees in SOUND FOUNDATIONS are fictitious.

1. How many quality points do you have? 1. __________

2. **ATTENDANCE:**

   a. Pick an "Away Concert" card. Where are you performing? 2a. __________

   b. What is the concert hall's capacity? (See the chart.) b. __________

   c. Divide your quality points (No. 1) by 800 and round to the nearest hundredth. c. __________

   d. Find your attendance by multiplying the answer in No. 2c by the capacity (No. 2b). d. __________

3. **TICKET PRICE:**

Divide your quality points by 40 and round your answer to the nearest 50 cents. (This is the cost of one ticket.) How much does one ticket cost? 3. ____________

4. Each person attending your concert (No. 2d) must buy a ticket. (See No. 3 for the ticket price.) What is your income from ticket sales? 4. ____________

5. What is your away concert booking fee*? (See the chart.) 5. ____________

6. **GAS EXPENSES:**

   a. How far away is your concert hall? (See the chart.) 6a. ____________

   b. What is the round-trip distance? b. ____________

   c. If your truck gets 20 m.p.g., how many gallons of gas will you need? (Round to two decimal places.) c. ____________

   d. At a price of $1.30 per gallon, what is your total gas expense? d. ____________

7. **HOTEL EXPENSES:**

   a. You are taking 12 people (roadies, band, agent, and so forth) on the trip. It costs $33 per person for one night in the hotel. What is the total hotel cost? 7a. ____________

   b. What is the tax rate in your concert state? (See the chart.) b. ____________

   c. What is the amount of tax on your hotel bill? c. ____________

   d. What are your total hotel expenses? d. ____________

8. What are your total expenses for gas, hotel, and booking the hall? 8. ____________

9. What is the total profit after you pay all the expenses?

9. ________________

10. Pick a "Sharps and Flats" card. Record the card information. 10. ________________

11. What is your adjusted profit? (Round to the nearest dollar.) 11. [ ]

12. The manager gets 12% of the adjusted profit for the away concerts. What is your share?

Manager________________________ Date________________________

MILESTONE 9
FORM O

AWAY CONCERT

AWAY CONCERT HALL	LOCATION	CAPACITY	DISTANCE FROM NEW YORK CITY	BOOKING FEE*	STATE TAX RATE
Atlantic City Auditorium	Atlantic City, NJ	40,000	150 miles	$19,000	6%
Philadelphia Convention Hall	Philadelphia, PA	9,500	100 miles	$ 9,000	6%
Veteran's Coliseum	New Haven, CT	8,700	80 miles	$ 8,000	7.5%
Hershey Park Arena	Hershey, PA	7,200	160 miles	$ 7,000	6%
J.F.K. Center	Washington, DC	2,700	230 miles	$ 2,100	6%

*The booking fees in SOUND FOUNDATIONS are fictitious.

1. How many quality points do you have? 1. ____________

2. **ATTENDANCE:**

   a. Pick an "Away Concert" card. Where are you performing? 2a. ____________

   b. What is the concert hall's capacity? (See the chart.) b. ____________

   c. Divide your quality points (No. 1) by 800 and round to the nearest hundredth. c. ____________

   d. Find your attendance by multiplying the answer in No. 2c by the capacity (No. 2b). d. ____________

3. **TICKET PRICE:**

   Divide your quality points by 40 and round your answer to the nearest 50 cents. (This is the cost of one ticket.) How much does one ticket cost? 3. ____________

4. Each person attending your concert (No. 2d) must buy a ticket. (See No. 3 for the ticket price.) What is your income from ticket sales? 4. ____________

5. What is your away concert booking fee*? (See the chart.) 5. ____________

6. **GAS EXPENSES:**

   a. How far away is your concert hall? (See the chart.) 6a. ____________

   b. What is the round-trip distance? b. ____________

c. If your truck gets 20 m.p.g., how many gallons of gas will you need? (Round to two decimal places.) c. ____________

d. At a price of $1.30 per gallon, what is your total gas expense? d. ____________

7. **HOTEL EXPENSES:**

a. You are taking 12 people (roadies, band, agent, and so forth) on the trip. It costs $33 per person for one night in the hotel. What is the total hotel cost? 7a. ____________

b. What is the tax rate in your concert state? (See the chart.) b. ____________

c. What is the amount of tax on your hotel bill? c. ____________

d. What are your total hotel expenses? d. ____________

8. What are your total expenses for gas, hotel, and booking the hall? 8. ____________

9. What is the total profit after you pay all the expenses? 9. ____________

10. Pick a "Sharps and Flats" card. Record the card information. 10. ____________

11. What is your adjusted profit? (Round to the nearest dollar.) 11. ____________

12. The manager gets 12% of the adjusted profit for the away concerts. What is your share? 12. ____________

13. Deposit the manager's share in your Check Register.

SEE ACTIVITY 41 FOR PRACTICE VERSION

Manager____________________ Date____________________

MILESTONE 9
FORM O

AWAY CONCERT

AWAY CONCERT HALL	LOCATION	CAPACITY	DISTANCE FROM NEW YORK CITY	BOOKING FEE*	STATE TAX RATE
Atlantic City Auditorium	Atlantic City, NJ	40,000	150 miles	$19,000	6%
Philadelphia Convention Hall	Philadelphia, PA	9,500	100 miles	$ 9,000	6%
Veteran's Coliseum	New Haven, CT	8,700	80 miles	$ 8,000	7.5%
Hershey Park Arena	Hershey, PA	7,200	160 miles	$ 7,000	6%
J. F. K. Center	Washington, DC	2,700	230 miles	$ 2,100	6%

*The booking fees in SOUND FOUNDATIONS are fictitious.

1. How many quality points do you have? 1. __________

2. **ATTENDANCE:**

   a. Pick an "Away Concert" card. Where are you performing? 2a. __________

   b. What is the concert hall's capacity? (See the chart.) b. __________

   c. Divide your quality points (No. 1) by 800 and round to the nearest hundredth. c. __________

   d. Find your attendance by multiplying the answer in No. 2c by the capacity (No. 2b). d. __________

3. **TICKET PRICE:**

   Divide your quality points by 40 and round your answer to the nearest 50 cents. (This is the cost of one ticket.) How much does one ticket cost? 3. __________

4. Each person attending your concert (No. 2d) must buy a ticket. (See No. 3 for the ticket price.) What is your income from ticket sales? 4. __________

5. What is your away concert booking fee*? (See the chart.) 5. __________

6. **GAS EXPENSES:**

   a. How far away is your concert hall? (See the chart.) 6a. __________

   b. What is the round-trip distance? b. __________

c. If your truck gets 20 m.p.g., how many gallons of gas will you need? (Round to two decimal places.) c. ____________

d. At a price of $1.30 per gallon, what is your total gas expense? d. ____________

7. **HOTEL EXPENSES:**

a. You are taking 12 people (roadies, band, agent, and so forth) on the trip. It costs $33 per person for one night in the hotel. What is the total hotel cost? 7a. ____________

b. What is the tax rate in your concert state? (See the chart.) b. ____________

c. What is the amount of tax on your hotel bill? c. ____________

d. What are your total hotel expenses? d. ____________

8. What are your total expenses for gas, hotel, and booking the hall? 8. ____________

9. What is the total profit after you pay all the expenses? 9. ____________

10. Pick a "Sharps and Flats" card. Record the card information. 10. ____________

11. What is your adjusted profit? (Round to the nearest dollar.) 11. [ ]

12. The manager gets 12% of the adjusted profit for the away concerts. What is your share? 12. ____________

13. Deposit the manager's share in your Check Register.

SEE ACTIVITY 41 FOR PRACTICE VERSION

Manager________________________________ Date________________________

MILESTONE 9
FORM O

AWAY CONCERT

AWAY CONCERT HALL	LOCATION	CAPACITY	DISTANCE FROM NEW YORK CITY	BOOKING FEE*	STATE TAX RATE
Atlantic City Auditorium	Atlantic City, NJ	40,000	150 miles	$19,000	6%
Philadelphia Convention Hall	Philadelphia, PA	9,500	100 miles	$ 9,000	6%
Veteran's Coliseum	New Haven, CT	8,700	80 miles	$ 8,000	7.5%
Hershey Park Arena	Hershey, PA	7,200	160 miles	$ 7,000	6%
J. F. K. Center	Washington, DC	2,700	230 miles	$ 2,100	6%

*The booking fees in SOUND FOUNDATIONS are fictitious.

1. How many quality points do you have? 1. ____________

2. **ATTENDANCE:**

   a. Pick an "Away Concert" card. Where are you performing? 2a. ____________

   b. What is the concert hall's capacity? (See the chart.) b. ____________

   c. Divide your quality points (No. 1) by 800 and round to the nearest hundredth. c. ____________

   d. Find your attendance by multiplying the answer in No. 2c by the capacity (No. 2b). d. ____________

3. **TICKET PRICE:**

   Divide your quality points by 40 and round your answer to the nearest 50 cents. (This is the cost of one ticket.) How much does one ticket cost? 3. ____________

4. Each person attending your concert (No. 2d) must buy a ticket. (See No. 3 for the ticket price.) What is your income from ticket sales? 4. ____________

5. What is your away concert booking fee*? (See the chart.) 5. ____________

6. **GAS EXPENSES:**

   a. How far away is your concert hall? (See the chart.) 6a. ____________

   b. What is the round-trip distance? b. ____________

c. If your truck gets 20 m.p.g., how many gallons of gas will you need? (Round to two decimal places.) c. ____________

d. At a price of $1.30 per gallon, what is your total gas expense? d. ____________

7. **HOTEL EXPENSES:**

a. You are taking 12 people (roadies, band, agent, and so forth) on the trip. It costs $33 per person for one night in the hotel. What is the total hotel cost? 7a. ____________

b. What is the tax rate in your concert state? (See the chart.) b. ____________

c. What is the amount of tax on your hotel bill? c. ____________

d. What are your total hotel expenses? d. ____________

8. What are your total expenses for gas, hotel, and booking the hall? 8. ____________

9. What is the total profit after you pay all the expenses? 9. ____________

10. Pick a "Sharps and Flats" card. Record the card information. 10. ____________

11. What is your adjusted profit? (Round to the nearest dollar.) 11. ____________

12. The manager gets 12% of the adjusted profit for the away concerts. What is your share? 12. ____________

13. Deposit the manager's share in your Check Register.

SEE ACTIVITY 41 FOR PRACTICE VERSION

Manager ______________________ Date ______________________

MILESTONE 9
FORM O

AWAY CONCERT

AWAY CONCERT HALL	LOCATION	CAPACITY	DISTANCE FROM NEW YORK CITY	BOOKING FEE*	STATE TAX RATE
Atlantic City Auditorium	Atlantic City, NJ	40,000	150 miles	$19,000	6%
Philadelphia Convention Hall	Philadelphia, PA	9,500	100 miles	$ 9,000	6%
Veteran's Coliseum	New Haven, CT	8,700	80 miles	$ 8,000	7.5%
Hershey Park Arena	Hershey, PA	7,200	160 miles	$ 7,000	6%
J. F. K. Center	Washington, DC	2,700	230 miles	$ 2,100	6%

*The booking fees in SOUND FOUNDATIONS are fictitious.

1. How many quality points do you have? 1. ______________

2. **ATTENDANCE:**

   a. Pick an "Away Concert" card. Where are you performing? 2a. ______________

   b. What is the concert hall's capacity? (See the chart.) b. ______________

   c. Divide your quality points (No. 1) by 800 and round to the nearest hundredth. c. ______________

   d. Find your attendance by multiplying the answer in No. 2c by the capacity (No. 2b). d. ______________

3. **TICKET PRICE:**

   Divide your quality points by 40 and round your answer to the nearest 50 cents. (This is the cost of one ticket.) How much does one ticket cost? 3. ______________

4. Each person attending your concert (No. 2d) must buy a ticket. (See No. 3 for the ticket price.) What is your income from ticket sales? 4. ______________

5. What is your away concert booking fee*? (See the chart.) 5. ______________

6. **GAS EXPENSES:**

   a. How far away is your concert hall? (See the chart.) 6a. ______________

   b. What is the round-trip distance? b. ______________

c. If your truck gets 20 m.p.g., how many gallons of gas will you need? (Round to two decimal places.) c. ____________

d. At a price of $1.30 per gallon, what is your total gas expense? d. ____________

7. **HOTEL EXPENSES:**

a. You are taking 12 people (roadies, band, agent, and so forth) on the trip. It costs $33 per person for one night in the hotel. What is the total hotel cost? 7a. ____________

b. What is the tax rate in your concert state? (See the chart.) b. ____________

c. What is the amount of tax on your hotel bill? c. ____________

d. What are your total hotel expenses? d. ____________

8. What are your total expenses for gas, hotel, and booking the hall? 8. ____________

9. What is the total profit after you pay all the expenses? 9. ____________

10. Pick a "Sharps and Flats" card. Record the card information. 10. ____________

11. What is your adjusted profit? (Round to the nearest dollar.) 11. ____________

12. The manager gets 12% of the adjusted profit for the away concerts. What is your share? 12. ____________

13. Deposit the manager's share in your Check Register.

SEE ACTIVITY 41 FOR PRACTICE VERSION

Manager________________________ Date________________________

MILESTONE 9
FORM O

AWAY CONCERT

AWAY CONCERT HALL	LOCATION	CAPACITY	DISTANCE FROM NEW YORK CITY	BOOKING FEE*	STATE TAX RATE
Atlantic City Auditorium	Atlantic City, NJ	40,000	150 miles	$19,000	6%
Philadelphia Convention Hall	Philadelphia, PA	9,500	100 miles	$ 9,000	6%
Veteran's Coliseum	New Haven, CT	8,700	80 miles	$ 8,000	7.5%
Hershey Park Arena	Hershey, PA	7,200	160 miles	$ 7,000	6%
J. F. K. Center	Washington, DC	2,700	230 miles	$ 2,100	6%

*The booking fees in SOUND FOUNDATIONS are fictitious.

1. How many quality points do you have? 1. ________

2. **ATTENDANCE:**

   a. Pick an "Away Concert" card. Where are you performing? 2a. ________

   b. What is the concert hall's capacity? (See the chart.) b. ________

   c. Divide your quality points (No. 1) by 800 and round to the nearest hundredth. c. ________

   d. Find your attendance by multiplying the answer in No. 2c by the capacity (No. 2b). d. ________

3. **TICKET PRICE:**

   Divide your quality points by 40 and round your answer to the nearest 50 cents. (This is the cost of one ticket.) How much does one ticket cost? 3. ________

4. Each person attending your concert (No. 2d) must buy a ticket. (See No. 3 for the ticket price.) What is your income from ticket sales? 4. ________

5. What is your away concert booking fee*? (See the chart.) 5. ________

6. **GAS EXPENSES:**

   a. How far away is your concert hall? (See the chart.) 6a. ________

   b. What is the round-trip distance? b. ________

c. If your truck gets 20 m.p.g., how many gallons of gas will you need? (Round to two decimal places.) c. ____________

d. At a price of $1.30 per gallon, what is your total gas expense? d. ____________

7. **HOTEL EXPENSES:**

a. You are taking 12 people (roadies, band, agent, and so forth) on the trip. It costs $33 per person for one night in the hotel. What is the total hotel cost? 7a. ____________

b. What is the tax rate in your concert state? (See the chart.) b. ____________

c. What is the amount of tax on your hotel bill? c. ____________

d. What are your total hotel expenses? d. ____________

8. What are your total expenses for gas, hotel, and booking the hall? 8. ____________

9. What is the total profit after you pay all the expenses? 9. ____________

10. Pick a "Sharps and Flats" card. Record the card information. 10. ____________

11. What is your adjusted profit? (Round to the nearest dollar.) 11. ____________

12. The manager gets 12% of the adjusted profit for the away concerts. What is your share? 12. ____________

13. Deposit the manager's share in your Check Register.

SEE ACTIVITY 41 FOR PRACTICE VERSION

Manager____________________ Date____________________

MILESTONE 9
FORM O

AWAY CONCERT

AWAY CONCERT HALL	LOCATION	CAPACITY	DISTANCE FROM NEW YORK CITY	BOOKING FEE*	STATE TAX RATE
Atlantic City Auditorium	Atlantic City, NJ	40,000	150 miles	$19,000	6%
Philadelphia Convention Hall	Philadelphia, PA	9,500	100 miles	$ 9,000	6%
Veteran's Coliseum	New Haven, CT	8,700	80 miles	$ 8,000	7.5%
Hershey Park Arena	Hershey, PA	7,200	160 miles	$ 7,000	6%
J. F. K. Center	Washington, DC	2,700	230 miles	$ 2,100	6%

*The booking fees in SOUND FOUNDATIONS are fictitious.

1. How many quality points do you have? 1. __________

2. **ATTENDANCE:**

   a. Pick an "Away Concert" card. Where are you performing? 2a. __________

   b. What is the concert hall's capacity? (See the chart.) b. __________

   c. Divide your quality points (No. 1) by 800 and round to the nearest hundredth. c. __________

   d. Find your attendance by multiplying the answer in No. 2c by the capacity (No. 2b). d. __________

3. **TICKET PRICE:**

   Divide your quality points by 40 and round your answer to the nearest 50 cents. (This is the cost of one ticket.) How much does one ticket cost? 3. __________

4. Each person attending your concert (No. 2d) must buy a ticket. (See No. 3 for the ticket price.) What is your income from ticket sales? 4. __________

5. What is your away concert booking fee*? (See the chart.) 5. __________

6. **GAS EXPENSES:**

   a. How far away is your concert hall? (See the chart.) 6a. __________

   b. What is the round-trip distance? b. __________

c. If your truck gets 20 m.p.g., how many gallons of gas will you need? (Round to two decimal places.) c. ______

d. At a price of $1.30 per gallon, what is your total gas expense? d. ______

7. **HOTEL EXPENSES:**

a. You are taking 12 people (roadies, band, agent, and so forth) on the trip. It costs $33 per person for one night in the hotel. What is the total hotel cost? 7a. ______

b. What is the tax rate in your concert state? (See the chart.) b. ______

c. What is the amount of tax on your hotel bill? c. ______

d. What are your total hotel expenses? d. ______

8. What are your total expenses for gas, hotel, and booking the hall? 8. ______

9. What is the total profit after you pay all the expenses? 9. ______

10. Pick a "Sharps and Flats" card. Record the card information. 10. ______

11. What is your adjusted profit? (Round to the nearest dollar.) 11. ______

12. The manager gets 12% of the adjusted profit for the away concerts. What is your share? 12. ______

13. Deposit the manager's share in your Check Register.

SEE ACTIVITY 41 FOR PRACTICE VERSION

Manager______________________________ Date______________________________

MILESTONE 9
FORM O

AWAY CONCERT

AWAY CONCERT HALL	LOCATION	CAPACITY	DISTANCE FROM NEW YORK CITY	BOOKING FEE*	STATE TAX RATE
Atlantic City Auditorium	Atlantic City, NJ	40,000	150 miles	$19,000	6%
Philadelphia Convention Hall	Philadelphia, PA	9,500	100 miles	$ 9,000	6%
Veteran's Coliseum	New Haven, CT	8,700	80 miles	$ 8,000	7.5%
Hershey Park Arena	Hershey, PA	7,200	160 miles	$ 7,000	6%
J. F. K. Center	Washington, DC	2,700	230 miles	$ 2,100	6%

*The booking fees in SOUND FOUNDATIONS are fictitious.

1. How many quality points do you have? 1. ____________

2. **ATTENDANCE:**

   a. Pick an "Away Concert" card. Where are you performing? 2a. ____________

   b. What is the concert hall's capacity? (See the chart.) b. ____________

   c. Divide your quality points (No. 1) by 800 and round to the nearest hundredth. c. ____________

   d. Find your attendance by multiplying the answer in No. 2c by the capacity (No. 2b). d. ____________

3. **TICKET PRICE:**

   Divide your quality points by 40 and round your answer to the nearest 50 cents. (This is the cost of one ticket.) How much does one ticket cost? 3. ____________

4. Each person attending your concert (No. 2d) must buy a ticket. (See No. 3 for the ticket price.) What is your income from ticket sales? 4. ____________

5. What is your away concert booking fee*? (See the chart.) 5. ____________

6. **GAS EXPENSES:**

   a. How far away is your concert hall? (See the chart.) 6a. ____________

   b. What is the round-trip distance? b. ____________

c. If your truck gets 20 m.p.g., how many gallons of gas will you need? (Round to two decimal places.) c. ________

d. At a price of $1.30 per gallon, what is your total gas expense? d. ________

7. **HOTEL EXPENSES:**

a. You are taking 12 people (roadies, band, agent, and so forth) on the trip. It costs $33 per person for one night in the hotel. What is the total hotel cost? 7a. ________

b. What is the tax rate in your concert state? (See the chart.) b. ________

c. What is the amount of tax on your hotel bill? c. ________

d. What are your total hotel expenses? d. ________

8. What are your total expenses for gas, hotel, and booking the hall? 8. ________

9. What is the total profit after you pay all the expenses? 9. ________

10. Pick a "Sharps and Flats" card. Record the card information. 10. ________

11. What is your adjusted profit? (Round to the nearest dollar.) 11. ________

12. The manager gets 12% of the adjusted profit for the away concerts. What is your share? 12. ________

13. Deposit the manager's share in your Check Register.

SEE ACTIVITY 41 FOR PRACTICE VERSION

Manager______________________ Date______________________

MILESTONE 9
FORM O

AWAY CONCERT

AWAY CONCERT HALL	LOCATION	CAPACITY	DISTANCE FROM NEW YORK CITY	BOOKING FEE*	STATE TAX RATE
Atlantic City Auditorium	Atlantic City, NJ	40,000	150 miles	$19,000	6%
Philadelphia Convention Hall	Philadelphia, PA	9,500	100 miles	$ 9,000	6%
Veteran's Coliseum	New Haven, CT	8,700	80 miles	$ 8,000	7.5%
Hershey Park Arena	Hershey, PA	7,200	160 miles	$ 7,000	6%
J. F. K. Center	Washington, DC	2,700	230 miles	$ 2,100	6%

*The booking fees in SOUND FOUNDATIONS are fictitious.

1. How many quality points do you have? 1. ______________

2. **ATTENDANCE:**

   a. Pick an "Away Concert" card. Where are you performing? 2a. ______________

   b. What is the concert hall's capacity? (See the chart.) b. ______________

   c. Divide your quality points (No. 1) by 800 and round to the nearest hundredth. c. ______________

   d. Find your attendance by multiplying the answer in No. 2c by the capacity (No. 2b). d. ______________

3. **TICKET PRICE:**

   Divide your quality points by 40 and round your answer to the nearest 50 cents. (This is the cost of one ticket.) How much does one ticket cost? 3. ______________

4. Each person attending your concert (No. 2d) must buy a ticket. (See No. 3 for the ticket price.) What is your income from ticket sales? 4. ______________

5. What is your away concert booking fee*? (See the chart.) 5. ______________

6. **GAS EXPENSES:**

   a. How far away is your concert hall? (See the chart.) 6a. ______________

   b. What is the round-trip distance? b. ______________

c. If your truck gets 20 m.p.g., how many gallons of gas will you need? (Round to two decimal places.) c. ____________

d. At a price of $1.30 per gallon, what is your total gas expense? d. ____________

7. **HOTEL EXPENSES:**

   a. You are taking 12 people (roadies, band, agent, and so forth) on the trip. It costs $33 per person for one night in the hotel. What is the total hotel cost? 7a. ____________

   b. What is the tax rate in your concert state? (See the chart.) b. ____________

   c. What is the amount of tax on your hotel bill? c. ____________

   d. What are your total hotel expenses? d. ____________

8. What are your total expenses for gas, hotel, and booking the hall? 8. ____________

9. What is the total profit after you pay all the expenses? 9. ____________

10. Pick a "Sharps and Flats" card. Record the card information. 10. ____________

11. What is your adjusted profit? (Round to the nearest dollar.) 11. [ ]

12. The manager gets 12% of the adjusted profit for the away concerts. What is your share? 12. ____________

13. Deposit the manager's share in your Check Register.

SEE ACTIVITY 41 FOR PRACTICE VERSION

Manager____________________ Date____________________

MILESTONE 9
FORM O

AWAY CONCERT

AWAY CONCERT HALL	LOCATION	CAPACITY	DISTANCE FROM NEW YORK CITY	BOOKING FEE*	STATE TAX RATE
Atlantic City Auditorium	Atlantic City, NJ	40,000	150 miles	$19,000	6%
Philadelphia Convention Hall	Philadelphia, PA	9,500	100 miles	$ 9,000	6%
Veteran's Coliseum	New Haven, CT	8,700	80 miles	$ 8,000	7.5%
Hershey Park Arena	Hershey, PA	7,200	160 miles	$ 7,000	6%
J. F. K. Center	Washington, DC	2,700	230 miles	$ 2,100	6%

*The booking fees in SOUND FOUNDATIONS are fictitious.

1. How many quality points do you have? 1. __________

2. **ATTENDANCE:**

   a. Pick an "Away Concert" card. Where are you performing? 2a. __________

   b. What is the concert hall's capacity? (See the chart.) b. __________

   c. Divide your quality points (No. 1) by 800 and round to the nearest hundredth. c. __________

   d. Find your attendance by multiplying the answer in No. 2c by the capacity (No. 2b). d. __________

3. **TICKET PRICE:**

   Divide your quality points by 40 and round your answer to the nearest 50 cents. (This is the cost of one ticket.) How much does one ticket cost? 3. __________

4. Each person attending your concert (No. 2d) must buy a ticket. (See No. 3 for the ticket price.) What is your income from ticket sales? 4. __________

5. What is your away concert booking fee*? (See the chart.) 5. __________

6. **GAS EXPENSES:**

   a. How far away is your concert hall? (See the chart.) 6a. __________

   b. What is the round-trip distance? b. __________

c. If your truck gets 20 m.p.g., how many gallons of gas will you need? (Round to two decimal places.) c. ________

d. At a price of $1.30 per gallon, what is your total gas expense? d. ________

7. **HOTEL EXPENSES:**

a. You are taking 12 people (roadies, band, agent, and so forth) on the trip. It costs $33 per person for one night in the hotel. What is the total hotel cost? 7a. ________

b. What is the tax rate in your concert state? (See the chart.) b. ________

c. What is the amount of tax on your hotel bill? c. ________

d. What are your total hotel expenses? d. ________

8. What are your total expenses for gas, hotel, and booking the hall? 8. ________

9. What is the total profit after you pay all the expenses? 9. ________

10. Pick a "Sharps and Flats" card. Record the card information. 10. ________

11. What is your adjusted profit? (Round to the nearest dollar.) 11. ________

12. The manager gets 12% of the adjusted profit for the away concerts. What is your share? 12. ________

13. Deposit the manager's share in your Check Register.

SEE ACTIVITY 41 FOR PRACTICE VERSION

Manager________________________ Date________________________

MILESTONE 9
FORM O

AWAY CONCERT

AWAY CONCERT HALL	LOCATION	CAPACITY	DISTANCE FROM NEW YORK CITY	BOOKING FEE*	STATE TAX RATE
Atlantic City Auditorium	Atlantic City, NJ	40,000	150 miles	$19,000	6%
Philadelphia Convention Hall	Philadelphia, PA	9,500	100 miles	$ 9,000	6%
Veteran's Coliseum	New Haven, CT	8,700	80 miles	$ 8,000	7.5%
Hershey Park Arena	Hershey, PA	7,200	160 miles	$ 7,000	6%
J. F. K. Center	Washington, DC	2,700	230 miles	$ 2,100	6%

*The booking fees in SOUND FOUNDATIONS are fictitious.

1. How many quality points do you have? 1. ____________

2. **ATTENDANCE:**

   a. Pick an "Away Concert" card. Where are you performing? 2a. ____________

   b. What is the concert hall's capacity? (See the chart.) b. ____________

   c. Divide your quality points (No. 1) by 800 and round to the nearest hundredth. c. ____________

   d. Find your attendance by multiplying the answer in No. 2c by the capacity (No. 2b). d. ____________

3. **TICKET PRICE:**

   Divide your quality points by 40 and round your answer to the nearest 50 cents. (This is the cost of one ticket.) How much does one ticket cost? 3. ____________

4. Each person attending your concert (No. 2d) must buy a ticket. (See No. 3 for the ticket price.) What is your income from ticket sales? 4. ____________

5. What is your away concert booking fee*? (See the chart.) 5. ____________

6. **GAS EXPENSES:**

   a. How far away is your concert hall? (See the chart.) 6a. ____________

   b. What is the round-trip distance? b. ____________

c. If your truck gets 20 m.p.g., how many gallons of gas will you need? (Round to two decimal places.) c. ______

d. At a price of $1.30 per gallon, what is your total gas expense? d. ______

7. **HOTEL EXPENSES:**

a. You are taking 12 people (roadies, band, agent, and so forth) on the trip. It costs $33 per person for one night in the hotel. What is the total hotel cost? 7a. ______

b. What is the tax rate in your concert state? (See the chart.) b. ______

c. What is the amount of tax on your hotel bill? c. ______

d. What are your total hotel expenses? d. ______

8. What are your total expenses for gas, hotel, and booking the hall? 8. ______

9. What is the total profit after you pay all the expenses? 9. ______

10. Pick a "Sharps and Flats" card. Record the card information. 10. ______

11. What is your adjusted profit? (Round to the nearest dollar.) 11. ______

12. The manager gets 12% of the adjusted profit for the away concerts. What is your share? 12. ______

13. Deposit the manager's share in your Check Register.

SEE ACTIVITY 41 FOR PRACTICE VERSION

MILESTONE 10

NATIONAL TOUR

Manager____________________ Date________________

ACTIVITY 42

THE METRIC SYSTEM

Foundations: Bands that tour outside the United States must be familiar with the metric system. Many things in the band's daily life will be affected by the metric system.

Directions: Answer each of the following.

1. Indicate with a checkmark (✓) in the blank to the right which of the following measurements would be affected if your band visited a country that used the metric system.

a.	The weight of your truck	1a. ________
b.	Your roadie's age	b. ________
c.	The volume of your speaker cabinets	c. ________
d.	The number of cents in a dollar	d. ________
e.	The distance from London to Rome	e. ________
f.	The time the sun rises	f. ________
g.	The height of the Eiffel Tower	g. ________
h.	A container of milk	h. ________
i.	The temperature in Iceland	i. ________
j.	The number of eggs in a dozen	j. ________
k.	Your shoe size	k. ________
l.	The speed limit	l. ________
m.	The temperature of boiling water	m. ________
n.	Cake recipes	n. ________
o.	The perimeter of a stage	o. ________
p.	20/20 vision	p. ________

2. Fill in the blanks.

a. 5.6 km = ________________ mm

b. 107 cg = ________________ mg

c. 18 L = ________________ daL

d. .147 kg = ________________ g

e. 3 g = ________________ cg

f. 16 m = ________________ mm

g. 53 L = ________________ mL

h. 12 kg = ________________ g

i. 7 g = ________________ kg

j. 22.6 mg = ________________ g

ACTIVITY 43

MAP SCALES

Foundations: Band managers must use map scales to figure out travel routes and gas expenses.

Directions: Answer the following questions.

Manager____________________ Date____________________

1. Your band is planning to travel from San Antonio to Houston. These two cities are 3 inches apart on your road map, and 1 inch represents 65 miles. How many miles is Houston from San Antonio?

   1. ____________________

2. A California road map uses a scale in which 1 inch represents 100 miles. On that map, Los Angeles is about $3\frac{3}{4}$ inches from San Francisco.

   a. About how many miles is the trip from Los Angeles to San Francisco? 2a. ____________________

   b. If your truck gets 30 m.p.g., how many gallons of gas would it use for the trip? b. ____________________

   c. At a price of $1.35 per gallon, what would your gas expenses be? c. ____________________

3. On a map, 1 inch represents 147 miles. Your band must travel from New Orleans to Miami. These two cities are 6 inches apart on the road map. What is the actual distance from New Orleans to Miami?

   3. ____________________

4. Your band is leaving from Toronto, Canada, and going to Chicago, Illinois. Every centimeter on the map represents 62.8 kilometers. The distance between the two cities measures 12.5 centimeters.

   a. What is the actual distance in kilometers? 4a. ____________________

   b. Approximately how long will the trip take if the driver averaged 78 kilometers per hour? Round your answer to the nearest whole number. b. ____________________

Manager________________________ Date____________________

ACTIVITY 44

COORDINATE GEOMETRY

Foundations: Coordinates are very helpful in map reading. They will help the roadies locate cities on the national tour.

Directions: Use the United States map to answer the questions.

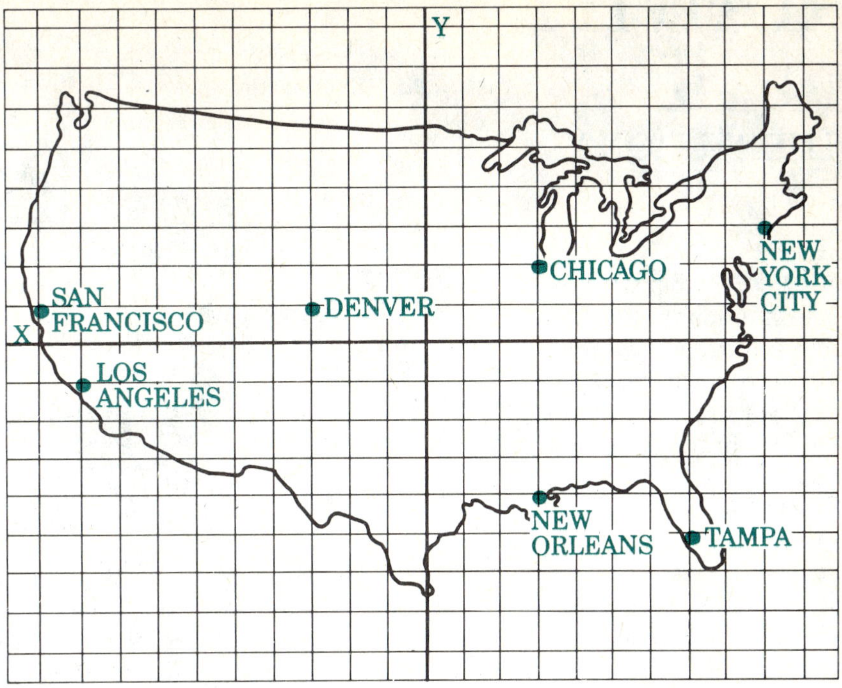

1. Give the coordinates of these cities:

   a. Tampa — 1a. ____________

   b. New Orleans — b. ____________

   c. New York City — c. ____________

   d. San Francisco — d. ____________

   e. Los Angeles — e. ____________

   f. Chicago — f. ____________

   g. Denver — g. ____________

2. Use the given coordinates to plot the following cities on the map:

   a. Washington, DC (8, 2)

   b. St. Louis (2, 0)

   c. Houston (1, −4)

   d. Las Vegas (−7, 0)

   e. Seattle (−9, 5)

   f. Nashville (5, −1)

   g. Cincinnati (5, 1)

ACTIVITY 45

THE NATIONAL TOUR

Foundations: The National Tour features concerts in four cities. There are travel days between each concert. Milestone 10/Form P – National Tour forms are similar to Milestone 9/Form O – Away Concert forms except that the band travels by plane, the hotel is booked for *two* nights, and the truck is not making *round* trips.

Directions: Alejandro's band is taking Tour 3 – Great Lakes Tour. The four band members are traveling from Detroit to play at Wrigley Field in Chicago. Alejandro's band has 700 quality points and will earn an extra $9,000 from a live radio concert (Sharps and Flats). Complete a Milestone 10/Form P – National Tour form for Alejandro.

Manager__________________________ Date____________________

PRACTICE VERSION

MILESTONE 10
FORM P

NATIONAL TOUR

1. How many quality points do you have? 1. ________________

2. What is your National Tour number? 2. ________________

3. **ATTENDANCE:**

   a. In what city are you performing for *this* Milestone 10/Form P – National Tour form? (See the information sheet.) 3a. ________________

   b. Where will your concert be held? (See the information sheet.) b. ________________

   c. What is the concert hall's capacity? (See the information sheet.) c. ________________

   d. Divide your quality points by 800. (Round to two decimal places.) d. ________________

   e. Find your attendance by multiplying the answer in No. 3d by the hall's capacity (No. 3c). e. ________________

4. **TICKET PRICE:**

   Divide your quality points by 40 and round your answer to the nearest 50 cents. This is the price of one ticket. How much does one ticket cost? 4. ________________

5. Each person attending your concert (No. 3e) must buy a ticket. (See No. 4 for the ticket price.) What is your income from the ticket sales? 5. ________________

6. What is the booking fee*? (See the information sheet.) 6. ____________________

7. **GAS EXPENSES:**

a. You are traveling from ____________________ to ____________________. (See the information sheet.)

b. What is the distance between the two cities? (See 7a and the information sheet.) 7b. ____________________

c. If your truck gets 20 m.p.g., how many gallons of gas will you need? (Round to two decimal places.) c. ____________________

d. Round your answer in No. 7c to the nearest gallon. d. ____________________

e. At a price of $1.30 per gallon, what is your total gas expense? e. ____________________

8. **HOTEL EXPENSES:**

a. You are taking 12 people (roadies, band, agent, and so forth) on the trip. It costs $33 per person for one night in the hotel. What is the total hotel cost for *two* nights? 8a. ____________________

b. What is the tax rate where your concert is held? (See the information sheet.) b. ____________________

c. What is the amount of tax on your hotel bill? c. ____________________

d. What are your total hotel expenses? d. ____________________

9. **AIRFARES:**

a. You are traveling from ____________________ to ____________________. (See No. 7a.)

b. How many people are in your band? 9b. ____________________

c. How much is each airfare? (See the information sheet.) c. ____________________

d. What is the total airfare expense? d. ______________

10. What are your total expenses for booking the hall, gas, hotel, and airfares? 10. ______________

11. What is the total profit after you pay all the expenses? 11. ______________

12. Pick a "Sharps and Flats" card. Record the card information. 12. ______________

13. What is your adjusted profit? (Round to the nearest dollar.) 13. [ ]

14. The manager gets 19% of the adjusted profit for each National Tour concert. What is your share? 14. ______________

*The booking fees in SOUND FOUNDATIONS are fictitious.

NATIONAL TOUR INFORMATION SHEET

TOUR	AIR-FARE	MILE-AGE	FLIGHT ITINERARY	CONCERT HALL	CONCERT HALL CAPACITY	BOOK-ING FEE	STATE SALES TAX RATE
1	$311	2,988	New York to Vancouver	Pacific Coliseum	15,500	$15,000	6.5%
	$ 60	144	Vancouver to Seattle	Kingdome	59,000	$49,000	6.5%
	$140	808	Seattle to San Francisco	Cow Palace	14,500	$13,000	4.75%
	$100	414	San Francisco to Los Angeles	Dodger Stadium	56,000	$60,000	4.75%
2	$180	1,758	New York to Denver	Mile High Stadium	57,200	$49,000	3%
	$175	499	Denver to Salt Lake City	Salt Palace	12,200	$11,000	4.6%
	$140	469	Salt Lake City to Las Vegas	Club Circuit	32,400	$31,000	5.75%
	$150	285	Las Vegas to Phoenix	Veterans Coliseum	13,200	$ 9,000	5%
3	$180	516	New York to Toronto	Maple Leaf Gardens	17,000	$11,000	5%
	$110	240	Toronto to Detroit	Olympia Stadium	16,200	$16,000	4%
	$100	275	Detroit to Chicago	Wrigley Field	29,600	$26,000	5%
	$100	289	Chicago to St. Louis	Busch Stadium	50,100	$51,000	4.2%
4	$170	872	New York to Nashville	Grand Ole Opry	3,500	$ 4,000	5.5%
	$150	780	Nashville to Houston	Astrodome	45,000	$45,000	4%
	$ 95	352	Houston to New Orleans	Superdome	26,300	$30,000	4%
	$170	860	New Orleans to Miami	Orange Bowl	58,000	$50,000	5%
5	$ 95	208	New York to Boston	Fenway Park	33,400	$35,000	5%
	$105	310	Boston to Montreal	Olympic Stadium	58,800	$41,000	5%
	$185	607	Montreal to Washington	Convention Hall	19,000	$19,000	6%
	$ 65	131	Washington to Philadelphia	Spectrum	18,200	$14,000	6%

*The booking fees in SOUND FOUNDATIONS are fictitious.

Manager____________________ Date____________________

MILESTONE 10
FORM P

NATIONAL TOUR

1. How many quality points do you have? 1. __________

2. What is your National Tour number? 2. __________

3. **ATTENDANCE:**

   a. In what city are you performing for *this* Milestone 10/Form P – National Tour form? (See the information sheet.) 3a. __________

   b. Where will your concert be held? (See the information sheet.) b. __________

   c. What is the concert hall's capacity? (See the information sheet.) c. __________

   d. Divide your quality points by 800. (Round to two decimal places.) d. __________

   e. Find your attendance by multiplying the answer in No. 3d by the hall's capacity (No. 3c). e. __________

4. **TICKET PRICE:**

   Divide your quality points by 40 and round your answer to the nearest 50 cents. This is the price of one ticket. How much does one ticket cost? 4. __________

5. Each person attending your concert (No. 3e) must buy a ticket. (See No. 4 for the ticket price.) What is your income from the ticket sales? 5. __________

6. What is the booking fee*? (See the information sheet.) 6. __________

7. **GAS EXPENSES:**

   a. You are traveling from __________ to __________. (See the information sheet.)

   b. What is the distance between the two cities? (See No. 7a and the information sheet.) 7b. __________

   c. If your truck gets 20 m.p.g., how many gallons of gas will you need? (Round to two decimal places.) c. __________

   d. Round your answer in No. 7c to the nearest gallon. d. __________

   e. At a price of $1.30 per gallon, what is your total gas expense? e. __________

Manager________________________ Date________________________

MILESTONE 10
FORM P

NATIONAL TOUR

1. How many quality points do you have? 1. ____________

2. What is your National Tour number? 2. ____________

3. **ATTENDANCE:**

   a. In what city are you performing for *this* Milestone 10/Form P – National Tour form? (See the information sheet.) 3a. ____________

   b. Where will your concert be held? (See the information sheet.) b. ____________

   c. What is the concert hall's capacity? (See the information sheet.) c. ____________

   d. Divide your quality points by 800. (Round to two decimal places.) d. ____________

   e. Find your attendance by multiplying the answer in No. 3d by the hall's capacity (No. 3c). e. ____________

4. **TICKET PRICE:**

   Divide your quality points by 40 and round your answer to the nearest 50 cents. This is the price of one ticket. How much does one ticket cost? 4. ____________

5. Each person attending your concert (No. 3e) must buy a ticket. (See No. 4 for the ticket price.) What is your income from the ticket sales? 5. ____________

6. What is the booking fee*? (See the information sheet.) 6. ____________

7. **GAS EXPENSES:**

   a. You are traveling from ____________ to ____________. (See the information sheet.)

   b. What is the distance between the two cities? (See No. 7a and the information sheet.) 7b. ____________

   c. If your truck gets 20 m.p.g., how many gallons of gas will you need? (Round to two decimal places.) c. ____________

   d. Round your answer in No. 7c to the nearest gallon. d. ____________

   e. At a price of $1.30 per gallon, what is your total gas expense? e. ____________

8. **HOTEL EXPENSES:**

a. You are taking 12 people (roadies, band, agent, and so forth) on the trip. It costs $33 per person for one night in the hotel. What is the total hotel cost for *two* nights? 8a. ________________

b. What is the tax rate where your concert is held? (See the information sheet.) b. ________________

c. What is the amount of tax on your hotel bill? c. ________________

d. What are your total hotel expenses? d. ________________

9. **AIRFARES:**

a. You are traveling from ________________ to ________________. (See No. 7a.)

b. How many people are in your band? 9b. ________________

c. How much is each airfare? (See the information sheet.) c. ________________

d. What is the total airfare expense? d. ________________

10. What are your total expenses for booking the hall, gas, hotel, and airfares? 10. ________________

11. What is the total profit after you pay all the expenses? 11. ________________

12. Pick a "Sharps and Flats" card. Record the card information. 12. ________________

13. What is your adjusted profit? (Round to the nearest dollar.) 13. [ ]

14. The manager gets 19% of the adjusted profit for each National Tour concert. What is your share? 14. ________________

15. Deposit the manager's share in your Check Register.

*The booking fees in SOUND FOUNDATIONS are fictitious.

SEE ACTIVITY 45 FOR PRACTICE VERSION

Manager____________________ Date____________________

MILESTONE 10
FORM P

NATIONAL TOUR

1. How many quality points do you have? 1. ________

2. What is your National Tour number? 2. ________

3. **ATTENDANCE:**

   a. In what city are you performing for *this* Milestone 10/Form P – National Tour form? (See the information sheet.) 3a. ________

   b. Where will your concert be held? (See the information sheet.) b. ________

   c. What is the concert hall's capacity? (See the information sheet.) c. ________

   d. Divide your quality points by 800. (Round to two decimal places.) d. ________

   e. Find your attendance by multiplying the answer in No. 3d by the hall's capacity (No. 3c). e. ________

4. **TICKET PRICE:**

   Divide your quality points by 40 and round your answer to the nearest 50 cents. This is the price of one ticket. How much does one ticket cost? 4. ________

5. Each person attending your concert (No. 3e) must buy a ticket. (See No. 4 for the ticket price.) What is your income from the ticket sales? 5. ________

6. What is the booking fee*? (See the information sheet.) 6. ________

7. **GAS EXPENSES:**

   a. You are traveling from ________ to ________. (See the information sheet.)

   b. What is the distance between the two cities? (See No. 7a and the information sheet.) 7b. ________

   c. If your truck gets 20 m.p.g., how many gallons of gas will you need? (Round to two decimal places.) c. ________

   d. Round your answer in No. 7c to the nearest gallon. d. ________

   e. At a price of $1.30 per gallon, what is your total gas expense? e. ________

8. **HOTEL EXPENSES:**

a. You are taking 12 people (roadies, band, agent, and so forth) on the trip. It costs $33 per person for one night in the hotel. What is the total hotel cost for *two* nights? 8a. ______________

b. What is the tax rate where your concert is held? (See the information sheet.) b. ______________

c. What is the amount of tax on your hotel bill? c. ______________

d. What are your total hotel expenses? d. ______________

9. **AIRFARES:**

a. You are traveling from ______________ to ______________. (See No. 7a.)

b. How many people are in your band? 9b. ______________

c. How much is each airfare? (See the information sheet.) c. ______________

d. What is the total airfare expense? d. ______________

10. What are your total expenses for booking the hall, gas, hotel, and airfares? 10. ______________

11. What is the total profit after you pay all the expenses? 11. ______________

12. Pick a "Sharps and Flats" card. Record the card information. 12. ______________

13. What is your adjusted profit? (Round to the nearest dollar.) 13. ______________

14. The manager gets 19% of the adjusted profit for each National Tour concert. What is your share? 14. ______________

15. Deposit the manager's share in your Check Register.

*The booking fees in SOUND FOUNDATIONS are fictitious.

SEE ACTIVITY 45 FOR PRACTICE VERSION

Manager____________________ Date____________________

MILESTONE 10
FORM P

NATIONAL TOUR

1. How many quality points do you have? 1. __________

2. What is your National Tour number? 2. __________

3. **ATTENDANCE:**

   a. In what city are you performing for *this* Milestone 10/Form P – National Tour form? (See the information sheet.) 3a. __________

   b. Where will your concert be held? (See the information sheet.) b. __________

   c. What is the concert hall's capacity? (See the information sheet.) c. __________

   d. Divide your quality points by 800. (Round to two decimal places.) d. __________

   e. Find your attendance by multiplying the answer in No. 3d by the hall's capacity (No. 3c). e. __________

4. **TICKET PRICE:**

   Divide your quality points by 40 and round your answer to the nearest 50 cents. This is the price of one ticket. How much does one ticket cost? 4. __________

5. Each person attending your concert (No. 3e) must buy a ticket. (See No. 4 for the ticket price.) What is your income from the ticket sales? 5. __________

6. What is the booking fee*? (See the information sheet.) 6. __________

7. **GAS EXPENSES:**

   a. You are traveling from __________ to __________. (See the information sheet.)

   b. What is the distance between the two cities? (See No. 7a and the information sheet.) 7b. __________

   c. If your truck gets 20 m.p.g., how many gallons of gas will you need? (Round to two decimal places.) c. __________

   d. Round your answer in No. 7c to the nearest gallon. d. __________

   e. At a price of $1.30 per gallon, what is your total gas expense? e. __________

8. **HOTEL EXPENSES:**

   a. You are taking 12 people (roadies, band, agent, and so forth) on the trip. It costs $33 per person for one night in the hotel. What is the total hotel cost for *two* nights? 8a. ____________

   b. What is the tax rate where your concert is held? (See the information sheet.) b. ____________

   c. What is the amount of tax on your hotel bill? c. ____________

   d. What are your total hotel expenses? d. ____________

9. **AIRFARES:**

   a. You are traveling from ____________ to ____________. (See No. 7a.)

   b. How many people are in your band? 9b. ____________

   c. How much is each airfare? (See the information sheet.) c. ____________

   d. What is the total airfare expense? d. ____________

10. What are your total expenses for booking the hall, gas, hotel, and airfares? 10. ____________

11. What is the total profit after you pay all the expenses? 11. ____________

12. Pick a "Sharps and Flats" card. Record the card information. 12. ____________

13. What is your adjusted profit? (Round to the nearest dollar.) 13. [ ]

14. The manager gets 19% of the adjusted profit for each National Tour concert. What is your share? 14. ____________

15. Deposit the manager's share in your Check Register.

*The booking fees in SOUND FOUNDATIONS are fictitious.

SEE ACTIVITY 45 FOR PRACTICE VERSION

8. **HOTEL EXPENSES:**

a. You are taking 12 people (roadies, band, agent, and so forth) on the trip. It costs $33 per person for one night in the hotel. What is the total hotel cost for *two* nights? 8a. ______________

b. What is the tax rate where your concert is held? (See the information sheet.) b. ______________

c. What is the amount of tax on your hotel bill? c. ______________

d. What are your total hotel expenses? d. ______________

9. **AIRFARES:**

a. You are traveling from ______________ to ______________. (See No. 7a.)

b. How many people are in your band? 9b. ______________

c. How much is each airfare? (See the information sheet.) c. ______________

d. What is the total airfare expense? d. ______________

10. What are your total expenses for booking the hall, gas, hotel, and airfares? 10. ______________

11. What is the total profit after you pay all the expenses? 11. ______________

12. Pick a "Sharps and Flats" card. Record the card information. 12. ______________

13. What is your adjusted profit? (Round to the nearest dollar.) 13. ______________

14. The manager gets 19% of the adjusted profit for each National Tour concert. What is your share? 14. ______________

15. Deposit the manager's share in your Check Register.

*The booking fees in SOUND FOUNDATIONS are fictitious.

SEE ACTIVITY 45 FOR PRACTICE VERSION